anand mahindra

प्रग्रह Pragraha

#one who inspires; leads

A Tweebook

A Compilation of Super 36 Tweets

Curator @ Dr. Prachee Javadekar

Design & Tech Credit : Dr. Aditee N. Vaidya

Popular Prakashan

www.popularprakashan.com

#Covid19

Israeli Tech

anand mahindra
@anandmahindra

Whenever I read about a scientific/technical breakthrough in Israel I tend to give that news a high credibility rating. Their ability to innovate rapidly is breathtaking. I hope we can work swiftly with them to import this critical capability

Israeli team has coronavirus test kit to test dozens of people at once
This new testing method allows for testing asymptomatic people suspected of having the virus, as well as for dozens of tests to be carried out at once, speedin...
jpost.com

12:47 PM · Mar 21, 2020 · Twitter for iPhone

1.7K Retweets **115** Quote Tweets **8.4K** Likes

Tweet Link

https://twitter.com/anandmahindra/status/1241262741194788864

Economic Impact of Covid19

anand mahindra
@anandmahindra

While we wait to see the economic impact of COVID-19, Invest India has launched the Business Immunity Platform to help businesses navigate these tough times. Kudos to the Invest India team from a proud board member
Investindia.gov.in/BIP

Invest India @investindia · Mar 21
Is #COVID_19 disrupting your business? We're here to help with our Business Community Platform -a comprehensive resource for investors & the business community to access relevant info about India's fight against #Coronavirus! Visit bit.ly/business_immun... NOW! #WarAgainstVirus

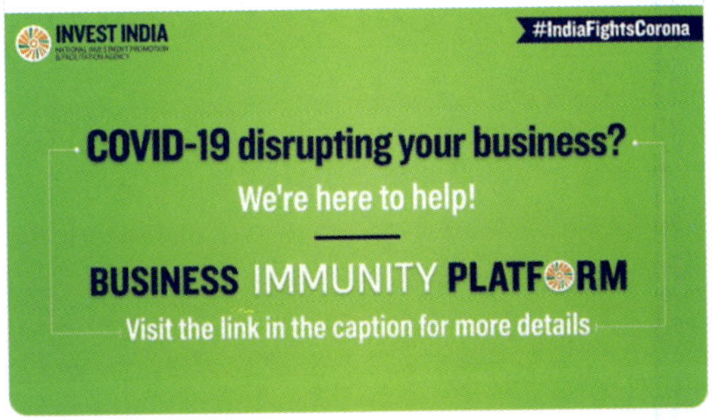

7:15 PM · Mar 22, 2020 · Twitter for iPhone

526 Retweets **25** Quote Tweets **2.9K** Likes

Tweet Link
https://twitter.com/anandmahindra/status/1241722670850297859

With great pleasure this Tweebook is presented to the readers.
This endevour comes to you at a much later time than was originally planned.

The original launching date was April 22, 2020
when Mr. Anand Mahindra's Twitter handle completed eleven long years.

The unfortunate arrival of the coronavirus pandemic delayed the arrival of the Tweebook. Yet, the silver lining is that it allowed valuable additions to the content with the inclusion of Covid19-related tweets in the print version of the book.

Mr. Mahindra notably continued to tweet prolifically and positively throughout the Covid19 triggered National Lockdown. An advantage of continued Lockdown would be that the readers may study the finer points of the Tweebook.

We hope this learning will remain with them for a long time to come.

Stay Home, Stay Safe!

#Covid19

Mahindra Lifespaces Ad

anand mahindra
@anandmahindra

I am, of course, waiting anxiously for life to get back to normal. But this ad from our team reminded me there's always a positive way of looking at our circumstances...

Mahindra Lifespaces @life_spaces · May 29
It's not just the time we spend at home, it's the lasting memories we build for years to come. So, let's cherish every moment with our loved ones because when routine life returns, these days will be missed the most.

#JoyfulHomecomings

11:24 AM · Jun 1, 2020 · Twitter for iPhone

280 Retweets **30** Quote Tweets **2.8K** Likes

Tweet Link
https://twitter.com/anandmahindra/status/1267333739882962944

Social Service

anand mahindra
@anandmahindra

To our team: You have opened up not only our kitchens, but our hearts. Thank you, from the bottom of my heart...

Pawan K Goenka @GoenkaPk · Apr 5
On clarion call from @PiyushGoyal mahindra opened its kitchen at 10 locations. We have supplied 50000 meals, 10000 rations this week. Making our kitchen infra available for others to use for up to 10000 meals a day.Please contact @shi_joshi. @MahindraRise

11:43 AM · Apr 5, 2020 · Twitter for iPad

3.5K Retweets **217** Quote Tweets **24.7K** Likes

Tweet Link
https://twitter.com/anandmahindra/status/1246682273774063616

#Covid19

Ventilator Manufacturing

anand mahindra @anandmahindra

So, so proud of our Kandivali & Igatpuri teams who confined themselves to the factories & without sleep produced this in 48hrs. With humility, we will seek guidance from specialists on the usefulness of the device. Whatever the outcome, they have shown India fights back...

12:38 PM · Mar 26, 2020 · Twitter for iPhone

5.1K Retweets **495** Quote Tweets **23K** Likes

Tweet Link
https://twitter.com/anandmahindra/status/1243072295167377408

Factory Reopening

anand mahindra @anandmahindra

Start well, start safe... Best wishes to you all, Team @MahindraAutoNA

Mahindra's Auburn Hills plant gets ready to restart vehicle assembly
"The crew was highly excited to start back up," Clarence Bringe said, "We want to start producing. That is our job. We want to produce."
clickondetroit.com

10:43 AM · May 14, 2020 · Twitter for iPad

233 Retweets **22** Quote Tweets **3.4K** Likes

Tweet Link
https://twitter.com/anandmahindra/status/1260800368063197185

Tweedeck

Ten years of regular tweeting by Mr. Anand Mahindra showed a clarity of thought and use of precise language covering a wide variety of topics.
I applied a unique method to analyze these tweets which I have named
'Theory Mapping'.

The analysis and interpretation of 36 tweets proved to be a revelatory experience.

~Dr. Prachee Javadekar, the Curator

anand mahindra
@anandmahindra

While wishing everyone a Happy & peaceful New Year, I also would like to say to those who find the peer pressure to make new resolutions & new beginnings somewhat oppressive: It's OK. Today is just another day. Keep doing what you love & the world will love what you do...

11:39 AM · Jan 1, 2020 · Twitter for iPhone

510 Retweets **7.4K** Likes

https://twitter.com/anandmahindra/status/1212254459415465985

The Tweebook Format

- Introduction to the category
- Academic theory or the essence of the category
- The protagonist's tweet along with the tweet link
- The curator's analysis of the tweet

Tweebook Birth Story

What is a Tweebook?

One knows of paperbacks and hard-bound books, even about e-books. These are formats which present writing of a considerable length.

A Tweebook, on the other hand, is a format which presents information in a shorter form. More specifically, it is a collation of content related to micro-blogging and its analysis.

One may ask, how can Twitter be a source of inspiration? Firstly, it is an extremely popular social media platform used to express opinions and share news. The use of hashtags has even become a valid form of grammar! Secondly, this site is used by people from all walks of life from all around the world. When certain personalities use this medium effectively, they are able to influence public opinion greatly.

This book is a direct result of the increasing use of the internet and digitization, which epitomizes this era. This is the first-ever endeavour at developing a Tweebook and the hope is that this format becomes as popular and readily available as other traditional book forms.

The Protagonist

Mr. Anand Mahindra

Padma Bhushan

#Iconic Industry Leader
#Committed to Values
#Transformational & Inspirational Leader

कर्तृत्वाद्यशसा प्रकाशयति यो स्वोद्योगक्षेत्रं महत्
नेतृत्वे सहभाजनाच्च विनयात्सर्वोच्चस्थानं जितम्
संदेशो समुदीरितोऽत्रभवता सिद्धिश्च सूत्रं तथा
संदेशाप्तकरंडकोऽयमखिलः सत्प्रेरणार्थं फलेत्

"He has illuminated the field of business and industry with his own achievements and reputation.
He has obtained the highest place in leadership with a sharing nature and humility.
His words give a formula for success.
This basket has the fruits of his wisdom which is inspirational."

@anandmahindra
Chairman, Mahindra Group
Mumbai, India

www.mahindra.com

Joined twitter : 22 April 2009

Followers	Tweets
7.19M	19.02K

Top 10 esteemed Influencers engaging with @anandmahindra Twitter handle

Source of Tweets

- Thought & Action
- Sensitivity & Empathy
- Inspiration & Awareness

Presentation of Tweets

- Use of Language
- Use of Multimedia
- Original & Retweets

Subjects Covered in Mr. Anand Mahindra's Tweets

- Leadership
- Branding
- Self Esteem
- Political Environment
- Domestic & Global Economic Scenario
- Policy Issues
- Technology & Innovation
- Women Empowerment
- Work-Life Balance
- Culture & Heritage
- Environment & Sustainability
- Sports

kaleidoscope

Padma Bhushan Mr. Anand Mahindra
Chairman, Mahindra Group

BUSINESS PROFILE

Decades	Enterprise	Companies	Countries	Employees
4+	US$20.7Bn	150+	100+	250K+

11 Sectors
22 Industries

- Steel
- Infrastructure
- Real Estate
- Aerospace
- Agriculture
- Automotive
- Boats
- Clean Energy
- Construction
- Defence
- Logistics
- Finance
- Insurance
- Power Backup
- Hospitality
- IT
- Infrastructure
- Retail

Mahindra
AIRVAN
VH-EKS
-EKS

Mahindra 575DI

Mahindra
AUTOMOTIVE NORTH AMERICA
ROXBOX

Mahindra
Rise.

susten
By Mahindra

Mahindra

Mahindra Group CSR activities

Literacy | Environment | Farmer Prosperity

Community-building | Women Empowerment

Club Mahindra

Mahindra

Tech Mahindra

mom & me
Mahindra Retail
Mom & Me is one-of-a-kind store for moms & kids.
view website

The Protagonist
#Mr. Anand Mahindra

Mr. Anand Mahindra is currently the Chairman of the Mahindra Group. Founded as a steel trading company in 1945, the Group today is a global federation of companies united by a common purpose - to enable people to **'Rise'**. On 26 January, 2020 Mr. Mahindra was awarded with Padma Bhushan- the third highest civilian award.

Mr. Mahindra was born on May 1, 1955 in Mumbai to Harish and Indira Mahindra. His father, a well-known industrialist, always emphasized the importance of using businesses to benefit society. Over the years, this principle has guided Mr. Mahindra's decisions and shaped the transformation of the Mahindra Group into one of the most admired Indian companies in the world.

Mr. Mahindra began his journey with the Mahindra Group as an Executive Assistant to the Finance Director of Mahindra Ugine Steel Company after receiving his MBA from Harvard Business School in 1981. In April 1997, his was appointed MD of Mahindra & Mahindra Ltd. Under his leadership, the Mahindra Group saw the successful acquisition of Ssangyong Motors, Reva Electric Car Company, Satyam Computer Services and several others. In August 2012, Mr. Mahindra was appointed Chairman of the Mahindra Group.

In 1996, he founded *Project Nanhi Kali*, an NGO to provide financial and academic support to underprivileged girls across India. Mr. Mahindra has also contributed to the advancement of the humanities. In November 2010, he donated $10 million to support the Harvard Humanities Center in honor of his mother.

the Driving Force

#Mr. Anand Mahindra

He has a keen interest in film-making and the visual arts, subjects he pursued as an undergraduate at Harvard College's Film Studies program. He is also a skilled photographer and enjoys listening to music.

Mr. Mahindra frequently shares his views and ideas on the Indian economy, sustainable urbanization, business and industry. He has also contributed to the book 'Re-imagining India' published by Mckinsey and Company.

Mr. Mahindra has regularly been noted as one of the most influential business leaders on Twitter. His Twitter handle - @anandmahindra, has over a million followers. In 2014, he was included in Fortune Magazine's list of 'World's 50 Greatest Leaders'.

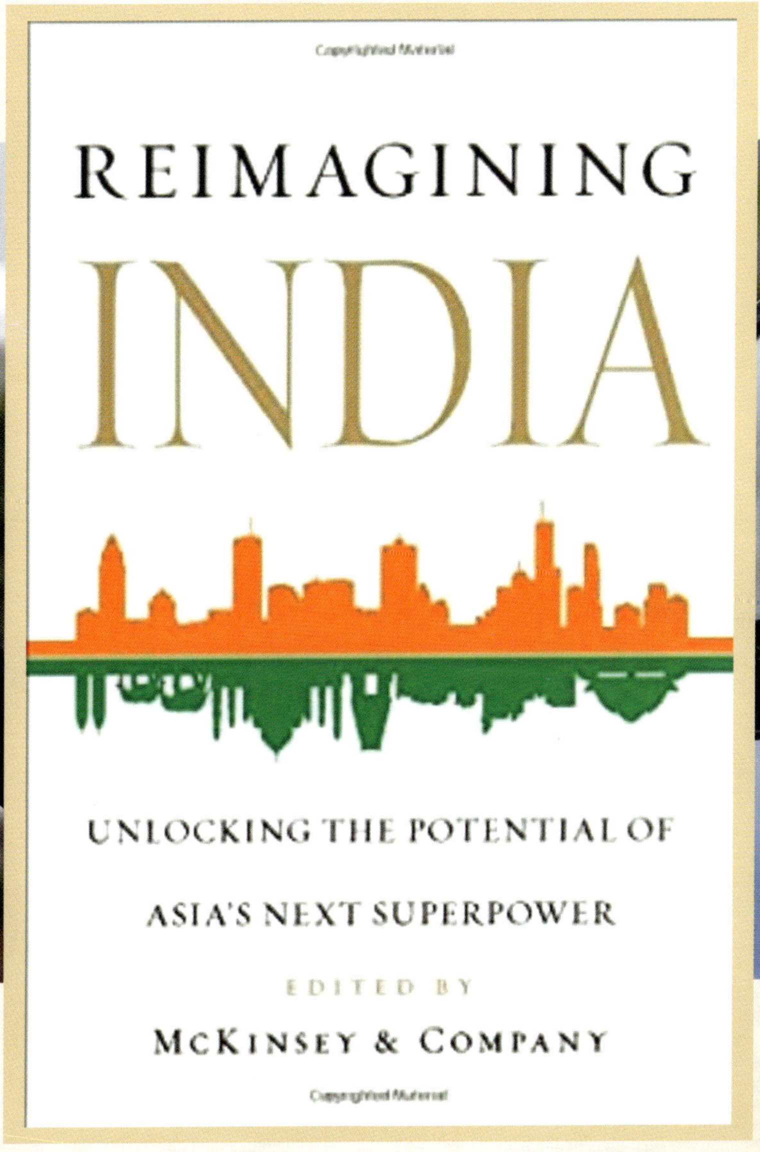
Copyrighted Material
REIMAGINING
INDIA
UNLOCKING THE POTENTIAL OF
ASIA'S NEXT SUPERPOWER
EDITED BY
McKINSEY & COMPANY
Copyrighted Material

The Curator

#Dr. Prachee Javadekar

A researcher and educationist, Dr. Prachee has been working as an academician and consultant for the last 40 years. She joined Indira Institute of Management as Director in 2000. Under her able leadership, this institute was ranked amongst top 20 management institutes in India. She then headed Javadekar Educational Consultancy Limited (JECL), and Parigha Research and Consultancy Services (PRCS), as a Director and CEO. She has authored and published more than 25 books and over 200 articles. Currently, she is an independent researcher in the field of documentation and publication.

Dr. Prachee's inclusive research approach has been a unique characteristic of every consultancy assignment she has handled till date. She has designed business consultancy offerings for a wide range of clients across the industry and corporate sector. This includes life-skills training, research case studies, corporate coffee table books and policy advocacy.

A deep understanding of the socio-economic dynamics of India has honed her thought processes. Further, her grasp of the policy perspective and challenges faced by the government and business sector is reflected in many of the unique services she has offered.

All the while a thorough professional, Dr. Prachee has a remarkable sensitivity towards those around her and that has allowed her to tap into the human experience in its entirety.

Tweebook, a first of its kind e-book

What inspired me to create this Tweebook which highlights Mr. Anand Mahindra's Transformational Leadership? When I revisit this question, I feel that the answer is simple. The passionate reader in me pushed me to do this!

Marathi being my mother-tongue, I have always been a voracious reader of its literature in all its forms. As an academician, I have also been exposed to writing by different authors of different genres, including contemporary writers. The list is long, with books ranging from cookery to chemistry, travel to trade, space to spirituality, education to entertainment and everything else! This has allowed me to internalize, process and express all that I have been exposed to, to my highest ability. In turn, this has sculpted and shaped the philosophy of my life.

Today, with technology evolving rapidly, the presentation techniques of literary material have also changed. New formats such as e-books have gained popularity. Now, new avenues of communication have also evolved including social media platforms, which encourage the micro-format of writing. These include Instagram, Twitter and Facebook, to name just a few. While following these, I have studied their reach and popularity.

Twitter has been especially attractive to me for its powerful micro-blogging capabilities. I have been following the Twitter feeds of many national and international personalities from diverse walks of life for many years. It has struck me that individuals, organizations and interest groups use and capitalize the power of this micro-blogging platform.

As my curiosity increased, I noticed that the latest technology affects these media on many levels. One of these was the ease of using the applications and posting material on them. Yet, what made me restless was that the quality of the content left a lot to be desired. A remarkable exception I noted was **@anandmahindra**. Mr. Mahindra is amongst the list of highly respected Indian business personalities to make a mark on Twitter and is on the top of my "Whom to Follow" list.

As an astute professional at the forefront of the business world, Mr. Mahindra has certainly been a personality to track. Before initiating this project, I had studied his professional journey through business magazines, news articles, interviews and speeches delivered at various business consortiums. His Twitter feed reinforced the personality that I already knew of and also brought to light many more of his facets.

The tweets had a strong intellectual base. This was especially true about an extra-ordinary tweet saluting the struggle of working women. In this tweet, Mr. Mahindra used a powerful visual with subtle words to pinpoint the reality of gender inequality. The likes, retweets and comments were insightful reading. Moreover, that this tweet gained national news coverage was also fascinating. This thought-provoking tweet proved to be the catalyst to this Tweebook!

Remarkably, ten years of regular tweeting showed a clarity of thought and the use of precise language covering a wide variety of topics. Mr. Mahindra's thoughts are rooted in Indian tradition yet his words speak to all generations. He not only takes criticism well, he also delivers it with wit and grace. This denotes a great positivity around his persona. His tweets seem to me to be an example of great modern writing.

While my emotional reaction to these tweets was understandable, still a Tweebook would have to be written on the basis of social-scientific research. As such, becoming a critical analyst was of prime importance.

Starting with vital statistics, as in the number of tweets, retweets and followers, data over a ten-year span was collated. The next phase was perusing the actual content of the tweets. It came to light that Mr. Mahindra has used Twitter to express thoughts about issues and experiences, appreciate people and personalities, describe his passions and hobbies, and comment on places and incidences. Along with my team, the facts, statistics, identities and phrases referred to in the tweets were further explored. As my research progressed, many previously unseen layers of Mr. Mahindra came to the fore. An image began to form which was that of a multi-dimensional individual. Soon, it was revealed that he was, in fact, the ideal example of a Transformational Leader.

At this point, it became clear that, distillation of the tweets to gain an essence was required to create a cohesive narrative.

For this, I led the team mentors in identifying and analyzing ten kinds of qualitative analysis formats. Tweets from August 2017 to August 2019 were studied and processed. Subsequently, six broad categories emerged in which the tweets would be grouped. The focus was now on the micro details of the tweets like the use of language, images and videos.

To analyze these tweets, I applied a unique method I have named **'Theory Mapping'**. This rigorous methodical approach helped unleash the academic basis of the tweets. A direct connection of a category of tweets with a theory base was identified. It became essential that the robust and vital academic worth of the content be unearthed. The analysis and interpretation of the tweets proved to be a revelatory experience for the entire team.

As a coherent body of information presented itself, it validated my initial feeling that Mr. Mahindra was not only a transformational leader but an inspirational leader as well. His professional and personal life revealed via the tweets portray him to be a humble, compassionate and dedicated leader. He has the rare quality of spearheading a large conglomerate with sharp business acumen, laced with great empathy and social sensitivity. These attributes are coupled with a proactive approach resulting in occasions of great philanthropy.

Finally, it was time to name the book. I thoughtfully chose a word that embodied all that had been gleaned from Mr. Mahindra's Twitter feed. The word ***Pragraha,*** meaning a person who leads and inspires, was a perfect fit. Choosing a Sanskrit word seemed apt as it reflected Mr. Mahindra's pride in India's cultural heritage. Continuing with this motif, original Sanskrit *shlokas* were developed and placed as the opening lines of each chapter.

The entire exercise of developing this Tweebook validated my admiration for Mr. Mahindra . In reality, it increased ten-fold!

I have a firm belief that this book is the need of the hour as it presents a cornucopia of life lessons to readers of all ages and walks of life across the globe. I hope the readers enjoy the experience of revelation this academic and creative endeavour will bring.

Albert Schweitzer
German polymath, theologian, organist, writer, humanitarian, philosopher, and physician.

"Example is not the MAIN THING in influencing others It is the ONLY THING"

@Being Anand Mahindra

#Business & Professionalism

व्यवसायश्च व्यावसायिकं कौशल्यम् च।

जिज्ञासा व्यवसायकार्यविषयादायामक्षेत्रे चिरम्
पण्यस्याध्ययनं विशेषगणनं सोपानविश्लेषणम्
सौजन्यं सुखदं विनम्रवचनैर्नेतृत्वमाधारवत्
सेवा सा परमा च कार्यनिपुणः सत्कर्मयोगी भवेत्

“Constant curiosity about the field of business and its various dimensions,
Deep study and analysis of business products,
Supportive leadership with courteous and humble words,
A person having such qualities serves society and is called ***Karmayogi.***”

#Business & Professionalism

The Essence of **#Business & Professionalism - By The Curator**

Anand Mahindra epitomises a committed business professional.

Professionalism conjures up a certain image of a person who has discipline, integrity and a commitment towards his work. A leader takes professionalism beyond the checklist of these traits.

A leader of the highest calibre embodies the values of humility, compassion and sportsmanship. Being inclusive, he transforms his team into a group of better workers and better human beings. This is because he is able to encourage excellence beyond developing a product or service.

All the while, Mr. Mahindra does not lose sight of his brand value and reputation. He is astute when creating, branding and positioning his company. His integrity is such that he appreciates self-competition rather than being in a race.

Theory Model for Business and Professionalism

Jim Collins is an American researcher author and consultant on business management and company sustainability. He believes that for a company's enduring presence in the marketplace, extraordinary leaders are required.

He, over a five-year period, did empirical research of 1000+ companies to distinguish 'good' companies from 'great' companies and authored the book **Good To Great.**[1]

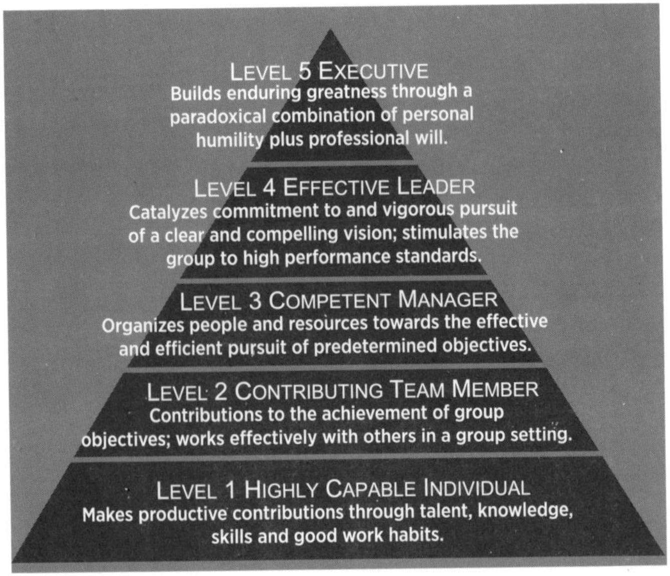

Collins found 11 companies that had made three extraordinary jumps over a period of 15 years and noticed that one of the key reasons was the contributions of the leaders of those companies. He named these leaders 'Level 5 Leaders'.

These leaders are predominantly a mixture of humility and resolute will which allow them to ensure their organizations become the best at what they do. Level 5 leaders are amazingly ambitious but their ambition is for a virtuous cause.

Jim Collins observes that Level 5 leaders have imbibed seven crucial principles - humility and resolve, selectivity and flexibility, realism and optimism, simplicity and purpose, discipline and measurement, purposeful technology and continuous improvement.

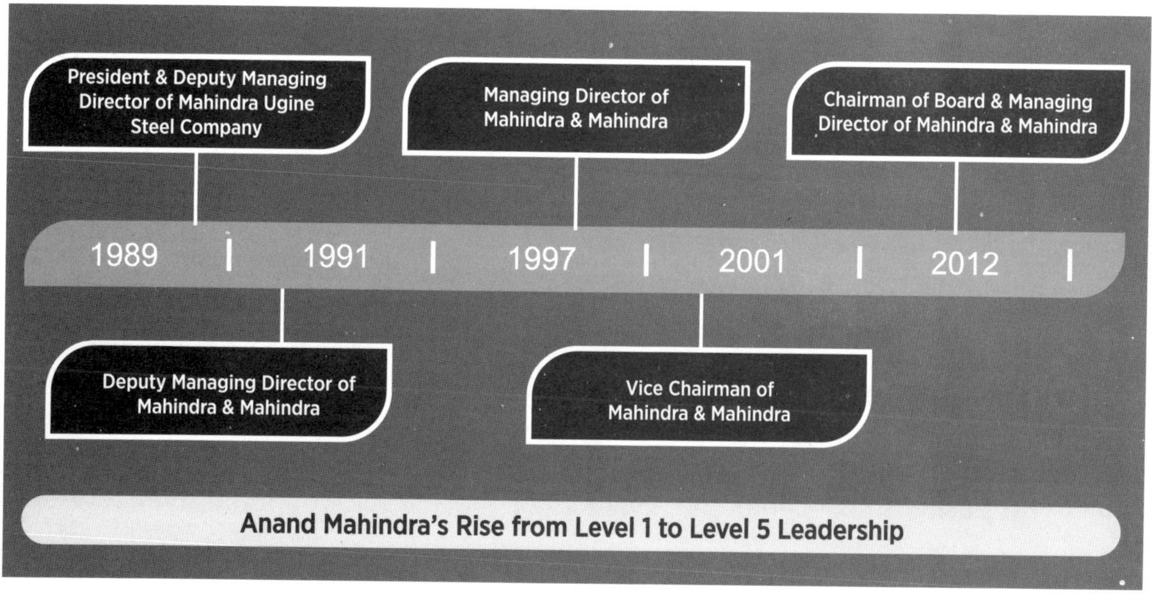

Anand Mahindra's Rise from Level 1 to Level 5 Leadership

The Situational Leadership model[2] evolves around four quadrants – Delegating, Supporting, Coaching and Directing. While greeting *twitterati* on Engineer's Day, he shows a spirited and youthful nature.[3] In a Caption Competition, Mr. Mahindra shared a picture of a bus with its own mirror image. He announced that the winning caption would get a die-cast scale model of a Mahindra vehicle. With such a competition, he gives the audience an opportunity to experience recognition.[4]

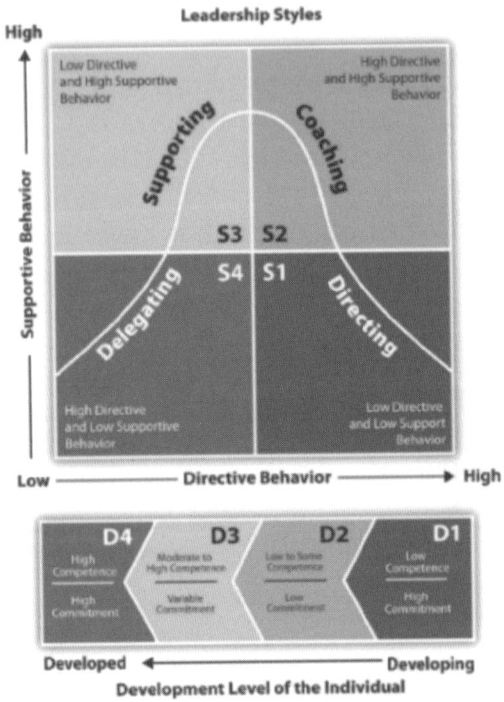

For a business, positioning its brand is very important. This brand positioning is created in the minds of the audience. Brands have to invest a lot of money in earning a brand position. In the tweet about how Mahindra vehicles were sent for support after the hurricane in the Bahamas, Mr. Mahindra reinstated his company's global brand positioning. This tweet also communicated support for a social cause which makes the brand socially responsible also.[5]

Good corporate citizenship is one of the values of the Mahindra Group. Actions give meaning to a brand. When a Twitter follower commented on the use of plastic bottles in Mahindra's Boardroom,[6] Mr. Mahindra immediately took action. There is no specific panacea for business challenges and a leader will reach a solution through efforts and vision.

Mr. Mahindra's career and tweets provide a lighted path for becoming a Level 5 leader. Mr. Mahindra has contributed to the nation's pride by making the Mahindra Group a US$20.7Bn enterprise, having a presence in more than 100 countries and with more than 2,50,000 employees.[7]

Super Six Parameters

1. Level 5 Leadership
2. Situational Leadership
3. Brand Personality
4. Motivational Leadership
5. Brand Positioning
6. Brand Meaning

1. https://www.jimcollins.com/concepts/level-five-leadership.html
2. http://atodorov.org/blog/2017/11/11/4-situational-leadership-styles/
3. https://twitter.com/anandmahindra/status/1173059814349574147
4. https://twitter.com/anandmahindra/status/1173821095583113216
5. https://twitter.com/anandmahindra/status/1174805152697585670
6. https://twitter.com/anandmahindra/status/1151114935813885953?ref_src=twsrc%5Etfw%7Ctwcamp%5Etweetembed%7Ctwterm%5E1151114935813885953&ref_url=https%3A%2F%2Fwww.hindustantimes.com%2Fit-s-viral%2Ftwitter-user-nudges-anand-mahindra-on-his-recent-post-his-response-is-epic%2Fstory-Q9rdxPYDDsiTv2uozdOq1J.html
7. https://www.mahindra.com/about-us

Level 5 Leader image source:
https://iwww.managementstudyguide.com/level-5-leadership.htm

#Business & Professionalism
@Being Anand Mahindra

Super Six

1

A Formula for Success

2

Roadmap for
the Financial Sector

3

Recognition of Strength

4
The Goal Setter
5
Sportsman's Spirit
Brand
6
Sustainable Mileage

#Business & Professionalism @Being Anand Mahindra

anand mahindra
@anandmahindra

Whenever I've been asked what the formula for success is, I have 1st stated that there is no 'silver bullet' or formula. But I have also said that I don't believe you can succeed in the long run if you do not possess a) Humility & b) Curiosity. In fact, one leads to the other...

Mahindra Rise @MahindraRise · Jul 3

Every innovation begins with questions.

Questions that challenge convention.

Only the curious few dare to ask these questions.

And that's why at Mahindra, we nurture curiosity in order to #FUTURise our businesses.

Do you #NurtureYourCuriosity?

Tweet Link
https://twitter.com/anandmahindra/status/1146385481245945858

1. A Formula for Success

From the Curator's Lens

According to Level 5 Leadership Theory, a Level 5 leader is one who is at the highest level of the leadership pyramid. Such a leader demonstrates great personal humility and strong professional will. This tweet portrays such a leader, who humbly accepts that there is no specific formula for success, but rather stresses that one must have a lifelong sense of wonder and inquisitiveness.

This is only one of many such tweets that prove Mr. Mahindra to be one who possesses these very rare qualities. For one who has reached such great heights in the business arena, he is also a most grounded human being.

#Business & Professionalism
@Being Anand Mahindra

anand mahindra
@anandmahindra

Yesterday, @udaykotak delivered the 25th Lalit Doshi lecture. He laid out precise prescriptions for the financial sector in order to fuel a $5T economy. If I were a policy maker looking for a roadmap for the financial sector I would have to look no further @nsitharaman

9:48 AM · Aug 6, 2019 · Twitter for iPhone

498 Retweets **3.6K** Likes

Tweet Link

https://twitter.com/anandmahindra/status/1158593133661175808

2. Roadmap for the Financial Sector

Mr. Anand Mahindra tweeted this to appreciate Mr. Uday Kotak and acknowledge his contribution to financial sector. This is an example of Situational Leadership Theory elaborating the different roles played by a leader. Situational Leadership Theory elaborates the different roles played by a leader. Here, more than one sector is connected by this tweet, making it that much richer and more noteworthy. A case in point, here a corporate leader has appreciated a financial leader's thoughts and connected them to the country's Finance Minister.

Showing a willingness to positively contribute towards the country's success is a significant leadership trait. The Mahindra Group plays a significant role in the Indian economy, as it recorded a revenue of approximately US $20.7 billion and retained about 200,000 employees worldwide in FY2019. This tweet reflects the very definition of a successful global leader who recognizes the expertise of another and shares that knowledge with due credit.

#Business & Professionalism @Being Anand Mahindra

anand mahindra
@anandmahindra

Frankly, this is the best testimonial for the vehicle's ruggedness & reliability...Better than any advertising that we could do...

Autocar Professional @autocarpro · Aug 31
Kolkata Police becomes 13th police force to deploy the compact Mahindra TUV300 for patrolling duties; total sales in India cross the 100,000-unit milestone. Full stats & facts bit.ly/2ZHXcKO

2:45 PM · Sep 1, 2019 · Twitter for iPad

289 Retweets **3.8K** Likes

Tweet Link
https://twitter.com/anandmahindra/status/1168090046211051520

3. Recognition of Strength

From the Curator's Lens

This tweet has shown appreciation for the Indian Police Force (IPF) as a whole. A correlation is sensed between the police force and a Mahindra SUV. The Mahindra TUV300 used by the IPF has a powerful mHawk100, 1.5 litre diesel engine with a 2-stage turbocharger. As such, it has passed the litmus test of "ruggedness" which embodies strength, speed and stability.

Communicated to the readers is the Mahindra brand personality. The images retweeted show a parking lot full of TUV300s and the police deputing them. The product and consumer are well-matched as both are symbols of strength. There is a subtle advertising of one's brand, and it is a straightforward example of heartfelt pride for both the police and the product.

#Business & Professionalism @Being Anand Mahindra

anand mahindra
@anandmahindra

When I joined M&M, one of my first responsibilities was the Tool&Die Shop at Nasik. I was always keen on them developing the capabilities to supply everything we needed in-house, but this has exceeded my greatest expectations...Thank you, Team Nasik, for making dreams a reality..

> **Vijay Kalra** @07Kalra · Jul 18
> Sharing a "Make in India" moment Pleased with Nasik Mahindra team for Conceptualizing, designing & building State of art " Robotic weld line". Normally would have gone to Japan/ Korea for same @anandmahindra @GoenkaPk @RajanWadhera1 @hsikka1 @autocarpro @MahindraRise @NITIAayog

6:29 PM · Jul 18, 2019 · Twitter for iPad

381 Retweets **3.8K** Likes

Tweet Link
https://twitter.com/anandmahindra/status/1151839004838359041

4. The Goal Setter

From the Curator's Lens

This tweet illustrates an important automation milestone achieved by The Mahindra Group's Nasik plant. The sense of joy and gratitude emanated, portrays a leader who sees the fulfillment of his dream of a full-fledged automated facility. That Mr. Mahindra chooses to mention **Make In India** (Government of India action plan introduced in 2014 as a part of a wider set of nation-building initiatives) reflects his sense of national pride.

Theory models prove that leadership involves the ability to motivate team members. Motivation is a continuous process over long periods of time as seen here. An effective leader starts the journey, regularly monitors it and takes it to its fruitful destination. Along with appreciating the company's growth, the tweet motivates the employees and stakeholders towards indigenous development.

#Business & Professionalism @Being Anand Mahindra

anand mahindra
@anandmahindra

We won't brag over this. It's an unfair contest.The Bolero's meant to negotiate such landscapes (seascapes?)But now you know why the Bolero's my favourite vehicle to drive. "Jaguar gets stuck in Mumbai floods: Mahindra Bolero drives through like a BOSS "

Jaguar gets stuck in Mumbai floods: Mahindra Bolero drives through like a BOSS...
A low slung sedan and flooded roads don't go together. A Jaguar luxury saloon found out the hard way earlier today, in a flooded underpass at Airoli, Navi ...
cartoq.com

11:12 PM · Sep 4, 2019 · Twitter for iPhone

4.6K Retweets **34.5K** Likes

Tweet Link

https://twitter.com/anandmahindra/status/1169304754373234689?lang=en

5. Sportsman's Spirit

The original tweet is self-explanatory as it is the news about the capabilities of a Mahindra vehicle. The article is gratifying as it reiterates the claims of the company about the **Bolero**. It has a tough, SUV build, high ground clearance, all-terrain capability and superior driving technology.

Mr. Mahindra is able to effectively employ this tweet to reflect his own confidence in his products. Also, he shows a sense of generosity by not comparing the two vehicles mentioned. In effect, the re-tweeting of the news item is an inadvertent brand positioning of the vehicle ahead of the competition.

#Business & Professionalism
@Being Anand Mahindra

anand mahindra
@anandmahindra

For those of you who asked for a better view of the car. This white #Battista is on display at the @PininfarinaSpA site. The blue was my favourite but I now think the white is even more stunning.. Watch when you have a leisurely couple of minutes... Don't fast forward. Savour it!

1:39 PM · Mar 6, 2019 · Twitter for iPhone

999 Retweets **6K** Likes

Tweet Link

https://twitter.com/anandmahindra/status/1103206045403570176

6. Sustainable Mileage

From the Curator's Lens

This tweet by Mr. Anand Mahindra is showing the image of the *Pininfarina Battista* at the 2019 Geneva Motor Show. Mahindra Racing Company's technology "Race to Road" is secured for this *Pininfarina Battista* vehicle.

Mr. Mahindra most excitedly posted not only the white model, but also the blue one on the day before. The state-of-the-art hand-built hyper-car boasts a 1,900hp/2,300 Nm torque which thrusts it to 100 km/h in less than two seconds and is capable of travelling 450 kms on one charge.

The quintessential question of a tweeter of *"Kitna Deti Hai?"* (What's the mileage?) was met with the witty reply, *"Sirji, electric hai..Shock deti hai!"* (Sir, it's an electric car...the mileage will surprise you!). Tweeting about this state-of-the-art e-vehicle places this trailblazer in the futuristic present. Likewise, it links the company with both the Indian ethos and global need for conserving the environment.

#Business & Professionalism
#Super Six Stats

	Formula for Success	Roadmap for the Financial Sector	Recognition of Strength
Retweets	161	508	293
Likes	1235	3609	3826
Comments	53	73	124

	The Goal Setter	Sportsman's Spirit	Sustainable Mileage
Retweets	386	4657	1010
Likes	3847	34840	6023
Comments	102	1900	500

"If your actions inspire others to dream more, learn more, do more and become more, you are a

LEADER"

John Quincy Adams
1767- 1848
6th U.S. President

@Being Anand Mahindra

#Inspirational Sharing

प्रेरणात्मकं सहभाजनम्।

उद्योगे सहभाजनेन हृदये संप्राप्यते प्रेरणा
आह्वानादवकाशलाभवरणाद्दुर्द्धर्षणं जायते
संघर्षे पथदर्शनं समुचितं धैर्यं समुत्साहनं
सिद्ध्यर्थं सहभाजनेन मनुजश्चैतत्समस्तं लभेत्

"Sharing in business brings inspiration in the heart. It creates challenges that give opportunities that make for great benefits and a spirit for fighting. It bestows proper guidance for the path to be taken, courage and enthusiasm in the times of struggle. If people are trying for success with sharing spirit, they obtain all of this."

#Inspirational Sharing

The Essence of **#Inspirational Sharing - By the Curator**

Anand Mahindra is indeed an inspirational leader as his individual values, beliefs and principles have crafted the Mahindra Group's ethical culture.

An inspirational leader becomes the catalyst for the transformation of his organization. The posts shared here represent two types of tweets. The first focuses on the people and events which inspired Mr. Mahindra and the second on how he inspires others. He is continually inspired by the greatness of others, be it tweets about PV Sindhu's or the ISRO scientists' achievements. He passes on the enthusiasm to be proactive and compassionate via tweets about delivery persons and disabled athletes.

This leader generates confidence in his followers who are inspired to achieve positive outcomes and, in the process, develop their own leadership capacity. His very persona and personality energizes those around him, be they the audience his speech is addressed to, his employees around the world or his followers on Twitter.

Mr. Mahindra spurs on every person he connects with, to flourish in a dynamic and complex world.

Theory Model for Inspirational Sharing

James Kouzes' and Barry Posner's research on leaders' traits suggests that leadership is not just a titular position, but a set of practices and behaviors. This is the Transformational Leadership Theory. The model they developed posits that an inspirational leader encourages followers to achieve extraordinary outcomes and, in the process, develop their own leadership capacity.

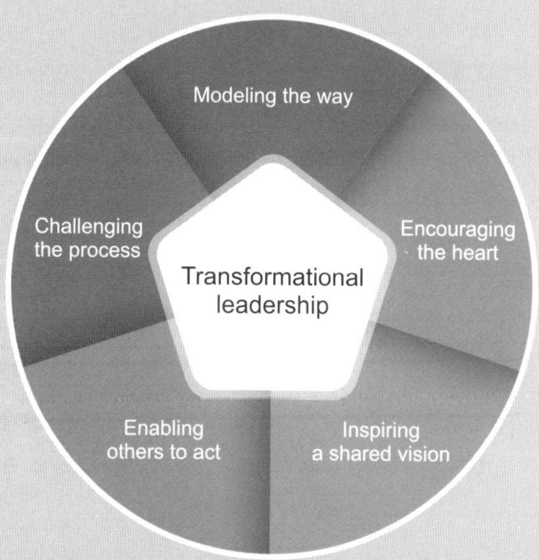

Transformational Leadership Theory Model

Transformational leaders foster employees' confidence to produce creative outcomes and sustain a competitive edge in an ever-changing business environment. They inspire, they act, they connect and they build the vision collectively with their followers.

Transformational leaders set an example and build commitment through daily acts that create progress and momentum. In order to model the way, the leaders need to have a philosophy, a set of high standards by which the organization is measured, a set of principles concerning the way people should be treated and the way goals should be pursued that make the organization unique and distinctive.

An important ability of a transformational leader is connecting emotionally with others. They listen carefully and respond rather than react to others. Furthermore, they are able to connect individual values to a larger shared vision. People often need encouragement and motivation to achieve the goals set by the organization. Successful leaders have high expectations for themselves and their employees. Their credibility is based on their record of achievements, dedication, and daily demonstrations of what and how things need to be done.[1] By influencing employee motivation, leaders attach rewards and recognition to job performance.

Transformational leaders play a special role in the celebrating of individual or group achievements. By this , leaders let people feel that they are a part of the group and a part of something significant.

Mr. Mahindra is such an inspirational leader, who has a set of high standards and principles by which the performance of his organization is measured. He has an ever-evolving philosophy by which he sets personal and professional goals. He uses social media, in this case Twitter, effectively when he tweets about topics that are of great relevance to the masses.

Mr. Mahindra's tweets bring to the fore instances of simplicity, dedication, positivity and humor as his personality has these very attributes. Examples of this include posts about the maverick fire-fighter Sarika Chimte[2], *Joote ka Doctor*[3] (The Shoe Doctor) or taking a stand for an employee on discrimination complaints[4]. He values an individual first, then the professional relationship.

He also has tweeted about other topics such as girl-child education and Indian agricultural sector, all of which serve to inspire the readers to expand their intellectual horizons. This creates a ripple effect that uplifts individuals and, in turn, society at large.

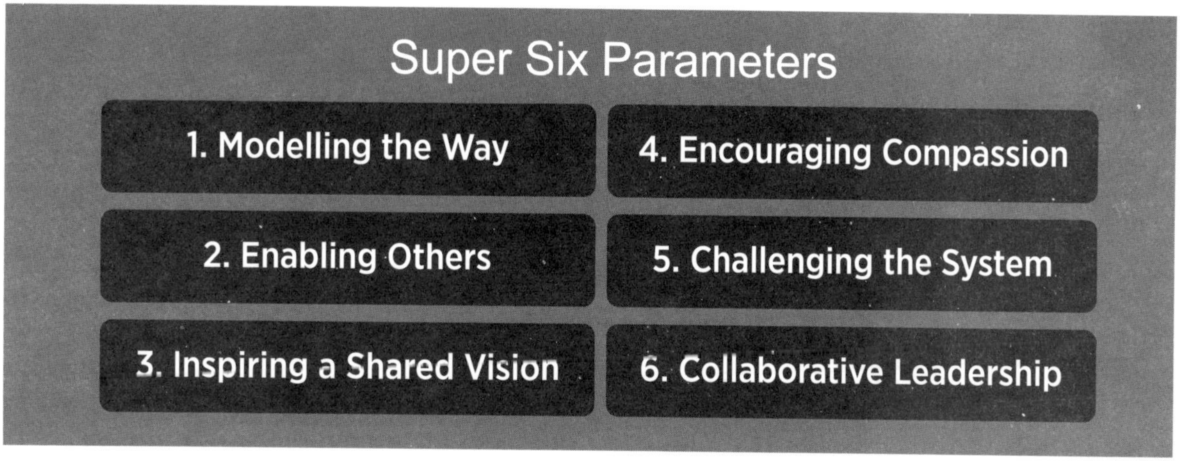

1. https://twitter.com/anandmahindra/status/1039427602556047361?lang=en
2. https://twitter.com/anandmahindra/status/994388254509223936
3. https://www.youtube.com/watch?v=G2OZSP6jgYY
4. https://www.moneycontrol.com/news/technology/former-tech-mahindra-employee-alleges-bigotry-anand-mahindra-and-gurnani-respond-2942201.html

Transformational Leadership Model image source:
https://www.americannursetoday.com/inspire-transformational-leadership/

#Inspirational Sharing
@Being Anand Mahindra
Super Six

1. The Champion's Way

2. Leading from the Front

3. The Three C's

4 Everyday Heroes

5 New Status Quo

6 Bucket List

#Inspirational Sharing @Being Anand Mahindra

anand mahindra
@anandmahindra

Brutal. I'm exhausted just watching this. But now there's no mystery about why she's the World Champ. A whole generation of budding Indian sportspersons will follow her lead & not shrink from the commitment required to get to the top...

9:17 PM · Aug 27, 2019 · Twitter for iPhone

16.4K Retweets **71.2K** Likes

Tweet Link
https://twitter.com/anandmahindra/status/1166376817265168385

1. The Champion's Way

From the Curator's Lens

By creating new ideas or recognizing and supporting old ones, leaders turn these ideas into actions. Mr. Mahindra is a keen supporter of physical fitness and sporting activities. Here, he tweets the video that went viral the day after Indian badminton player, PV Sindhu became the World Badminton Champion.

The video is a glimpse into the rigorous fitness regime PV Sindhu follows daily. To stay at No. 1, she is required to follow her routine with great consistency, perseverance and tenacity. Mr. Mahindra expresses a wish that aspiring sportspersons will be inspired by this testimony that there is no alternative to hard work.

#Inspirational Sharing
@Being Anand Mahindra

anand mahindra
@anandmahindra

We don't just want to be left out of farming's future. We want to play a role in shaping it..That's the logic behind this investment...

Autocar Professional @autocarpro · Jun 14
NEWS JUST IN: Mahindra & Mahindra acquires stake in Switzerland-based farmland analytics services company Gamaya; aims to develop and deploy next-gen farming capabilities such as precision agriculture and digital farming technologies bit.ly/2MN8HiX

1:12 PM · Jun 15, 2019 · Twitter for iPad

407 Retweets **3.3K** Likes

Tweet Link
https://twitter.com/anandmahindra/status/1139800252293079040

2. Leading from the Front

From the Curator's Lens

The quality of an inspirational leader is the ability to not only share a vision with a team, but also aid in its implementation. The Mahindra Group is committed to work for the betterment of Global agriculture. This is expressed best in one of their **Rise** brand stories, "Sowing the seeds of change and putting technology in the hands of every farmer."

This retweet shares the news of the acquisition of a stake in **Gamaya,** which is an innovative, futuristic tech firm in the field of agriculture. They operate in the precision farming sector and have created technologies and models for diagnosing farmland and crops. The Mahindra Group vision is to stay ahead of the times, so that the company pioneers advances in Global farming.

#Inspirational Sharing
@Being Anand Mahindra

anand mahindra
@anandmahindra

If you google the "Three C's" you'll find numerous quotes applicable to every field of activity. I like this one that I received today.
"Choice, Chance, Change. You must make the Choice to take the Chance if you want anything in life to Change."
Take some chances this week...

11:35 AM · Apr 8, 2019 · Twitter for iPad

1.6K Retweets **8.5K** Likes

Tweet Link
https://twitter.com/anandmahindra/status/1115133468470579200?lang=en

3. The Three C's

Transformational leaders passionately believe that they can make a difference around them and in the world at large. Mr. Mahindra leads by example in that he himself executes and implements the 3 C's. This *mantra* leads the readers on a life-altering process by making them think on a new plane.

His tweet carries a powerful suggestion for society at large to be innovative and creative. It attempts to impress upon them that taking risks is vital for growth and transformation. When a tweeter of this stature proposes such an action, it sparks the creation of a new generation of leaders.

#Inspirational Sharing
@Being Anand Mahindra

anand mahindra
@anandmahindra

Mumbai Airport runway shut, schools closed, train stations flooded, but the newspapers arrived in my house on time & dry! I have to salute those quiet, unsung heroes who brave torrential rain just so we can experience a 'normal day.' 🙏🙏🙏

Tweet Link
https://twitter.com/anandmahindra/status/1145912235752996865?lang=en

4. Everyday Heroes

From the Curator's Lens

During the first week of July 2019, heavy rains brought the city of Mumbai to a virtual standstill. Still, many parts of the city continued to receive essential services such as milk and newspaper deliveries.

Mr. Mahindra took this opportunity to salute the undeterred efforts of delivery persons who brave all kinds of weather to perform their duties. This tweet unveils a rare and exemplary belief in the dignity of labour. Genuine and profound thoughts serve to inspire and motivate the readers towards compassionate actions.

#Inspirational Sharing
@Being Anand Mahindra

anand mahindra
@anandmahindra

If @ProKabaddi had anything to do with inspiring the players in this video then, for me, this has to rank as the most satisfying outcome of the revival of the game. Thank you for sharing it...

Manoj Kumar @manoj_naandi · Mar 22
This video should warm the cockles of your heart. Who would have imagined the @ProKabaddi league envisioned and started by @anandmahindra would not only get people glued to the TV sets but also get all playing Kabbadi. Time to pitch it for Para Olympics @StarSportsIndia

10:04 AM · Mar 24, 2019 · Twitter for iPhone

706 Retweets **4.2K** Likes

Tweet Link
https://twitter.com/anandmahindra/status/1109674919057600512

5. A New Status Quo

From the Curator's Lens

The original tweet refers to the **Pro Kabaddi League**, the brainchild of Mr. Mahindra and Charu Sharma, started in 1994. 25 years of patience and perseverance has yielded the robust revival of the game of Kabaddi all over India.

For Mr. Mahindra, this relatively recent video of differently-abled players enjoying the game is a source of immense satisfaction. He feels fulfilled in the thought that the League he promoted has created a ripple effect that continues to touch the lives of many. Here, he appreciates the organizers of this tournament who have challenged the status quo which never visualized such an activity.

#Inspirational Sharing @Being Anand Mahindra

anand mahindra
@anandmahindra

Posted almost a year ago, but I had omitted to include it in my #OffTheBeatenTrack bucket list. Friday is a good day to post travel ideas!

Jaya Khare @jaya2004khare · Nov 4, 2018
Replying to @anandmahindra
Sir, Kipling's forest, where the coexistence of tigers, leopards, wild dogs, & sloth bears come together. am very excited to be able to showcase 2 u coexistence of humans & wildlife together in the real 'Jungle Book' setting. The national parks in #MP where one connects 2 nature.

8:30 AM · Sep 6, 2019 · Twitter for iPad

49 Retweets **963** Likes

Tweet Link
https://twitter.com/anandmahindra/status/1169807489682104320

6. Bucket List

From the Curator's Lens

This tweet was posted on a Friday, subtly encouraging followers to keep their gadgets aside and go on a trek on the weekend. The images in the original tweet are extremely attractive and enticing, lending a rich layer to the tweet.

When Mr. Mahindra tweets about his bucket list, he refers to spending the weekends in the proximity of nature. This is especially significant with respect to the corporate world where achieving a work-life balance is a real challenge. As such, a collaborative leader's suggestions tend to become well-received advice.

#Inspirational Sharing
#Super Six Stats

	The Champion's Way	Leading from the Front	The Three C's
Retweets	16600	414	113
Likes	71700	3281	429
Comments	1100	126	13

	Everyday Heroes	A New Status Quo	Bucket List
Retweets	770	711	50
Likes	5481	4227	975
Comments	74	55	34

Source-Shakti- power of every women Art by Mrinal Dutt

"Feminism or women empowerment isn't about making women stronger.
Women are already strong, it's about changing the way the world perceives that strength"

G. D. Anderson
Feminist and Activist

@Being Anand Mahindra

#Women Empowerment

स्त्रीशक्तिः हितकारिणी।

नारी शाश्वतपोषरक्षजननी वात्सल्यस्नेहांकुरी
नारी बोधकरी यशःसुखकरी सा ज्ञानबीजाक्षरी
नारी संतुलनाद्ददाति समतां त्राणाय दैत्यापहा
मुक्ता निर्णयनं करोति सकलं शक्तिभृता चित्कला

"Woman is Mother who always fosters and protects her children. She is a source of love and affection. She imparts insights, fame and happiness, and she has immortal seeds of knowledge. With balanced thinking, she teaches equality and becomes the destroyer of demons to protect the innocent. She is powerful and divine. She has the freedom of mind to take all her decisions."

The Essence of
#Women Empowerment - By the Curator

Anand Mahindra, a maverick business leader, stresses the importance and need of women empowerment efforts and has made it a goal of his company.

Women empowerment includes the action of raising the status of women through education, awareness, literacy, and training. When women are empowered, they are equipped to make decisions about their own lives. It is also the process by which women are able to redefine gender roles which gives them access to a wider range of social and economic opportunities.

The concepts of gender equality and gender equity are often hazy, but this is not so for Mr. Mahindra. He has clearly defined them just as he has great clarity about the necessity of striving for a work-life balance and the significance of parenting.

The Mahindra Group continually advocates women empowerment. Mr. Mahindra leads by example when he regularly tweets about the achievements of the women in the Mahindra Group, women in his family and women at large working across the diverse sectors.

Theory Model for Women Empowerment

Women Empowerment is generally defined as the process of acquiring the ability to make strategic life choices in a context where this ability was previously denied to them. The Three-Dimensional Model of women empowerment was developed by Marloes Anne Huis, Nina Hansen and Robert Lensink, experts in Women Studies from the University of Groningen, Netherlands.

The model presents three important components of the empowerment process, namely Resources, Agency and Achievements. Resources refer to material, human, and social expectations and allocations. Agency is the ability or sense of ability to define one's goals, act upon them, and decide on one's own strategic life outcomes.

Three Dimensional Model

Personal	Relational	Societal
Centre of control	Domestic injustice	Access to finance
Confidence	Bargaining power	Access to family decisions-mainly about marriage, pregnancy and abortion
Self esteem	Freedom of mobility	Participation at leadership roles
Self affectivity	Social network	Social acceptance
	Social capital	Freedom to select the social norm
	Participation level in problem solving at personal and collective level	Voicing power to large community platforms

Achievements include a variety of outcomes ranging from improved well-being to achieving equal representation of women in politics. According to the model, empowering women involves enabling them to exercise their individual choice in every sphere of life.

Mr. Mahindra, at an event of The Teenage Girls (TAG) Report[1] release said, "Giving safe working environment to women is not just something we alone have to do; everyone has to do it. I could get a little overly impassioned about the topic because I am the son of a feminist author, the husband of a career woman and my two daughters are professionals. So, my allegiance, I wear it on my sleeve."

Mr. Mahindra also actively endorses gender diversity and inclusivity in the workplace. His tweets portray his efforts of making workplaces safer for women.[2&3] Some tweets also highlight his appreciation for those women who have broken gender stereotypes.

Examples of such posts include tweets about Mahindra Rise initiatives such as *Prerna*,[4] for women farmers in India; *Nanhi Kali* [5] with #ListenToHer,[6] supporting girl-child education; Aarti the young female welder;[7] and Yogita Raghuvanshi the first Indian female truck driver.[8]

Mr. Mahindra has often expressed a deep admiration for his mother Indira Mahindra. He uses Twitter effectively when expressing how he was raised and how important good parenting is. He comes across as a humble son, supportive husband, encouraging father and caring grandparent.

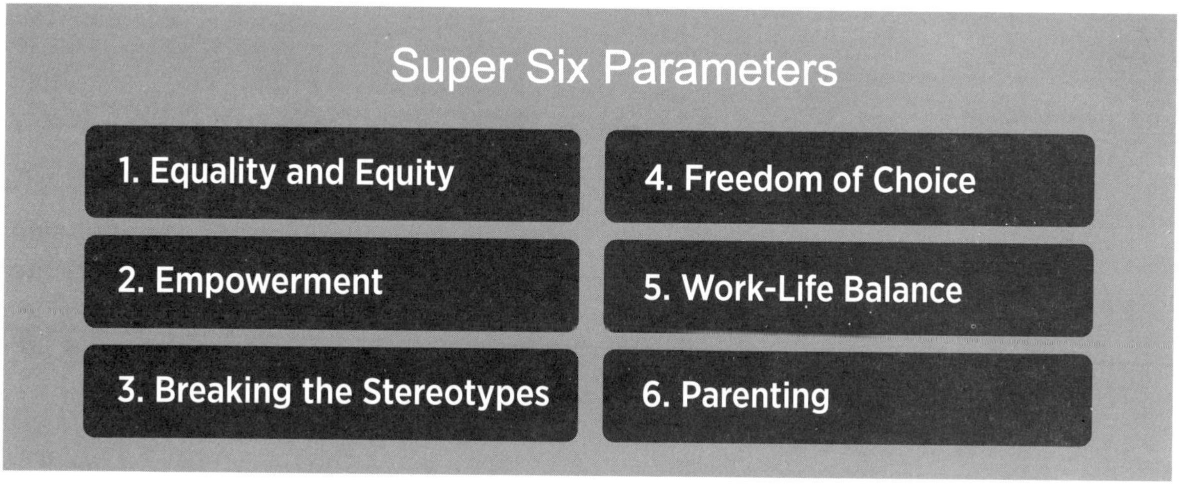

1.https://economictimes.indiatimes.com/magazines/panache/anand-mahindra-bats-for-more-women-in-boardrooms-says-10-pc-hike-in-girls-education-can-up-gdp-by-3-pc/articleshow/66525064.cms?from=mdr
2. https://twitter.com/anandmahindra/status/1183990136142348289?lang=en
3.https://economictimes.indiatimes.com/magazines/panache/anand-mahindra-has-zero-tolerance-for-sexual-predators-at-workplace-warns-employees-against-flouting-posh-norms/articleshow/71595300.cms
4. https://twitter.com/anandmahindra/status/1052123698025492480
5. https://www.nanhikali.org/
6. https://twitter.com/anandmahindra/status/1055474827828625408
7. https://twitter.com/anandmahindra/status/1028193301956976641
8. https://twitter.com/anandmahindra/status/1117759053701103617

3D Model image source:
https://pdfs.semanticscholar.org/f962/8f0b363facc89a1c0fcb45a8d755711b3e38.pdf?_ga=2.78045652.1946988370.1568361046-224415359.1568361046

#Women Empowerment
@Being Anand Mahindra

Super Six

1 A Level Playing Field

2 *Prerna*

3 It Can Be Done

4

Paper Tigress

5

Skewed Competition

6

Celebrating Mothers' Day

#Women Empowerment @Being Anand Mahindra

anand mahindra
@anandmahindra

Our Group Diversity Council made a presentation yesterday during which they used these images. A brilliant way of illustrating the vital difference between equality & equity; and why we need to create a level playing field for those who have been historically disadvantaged...

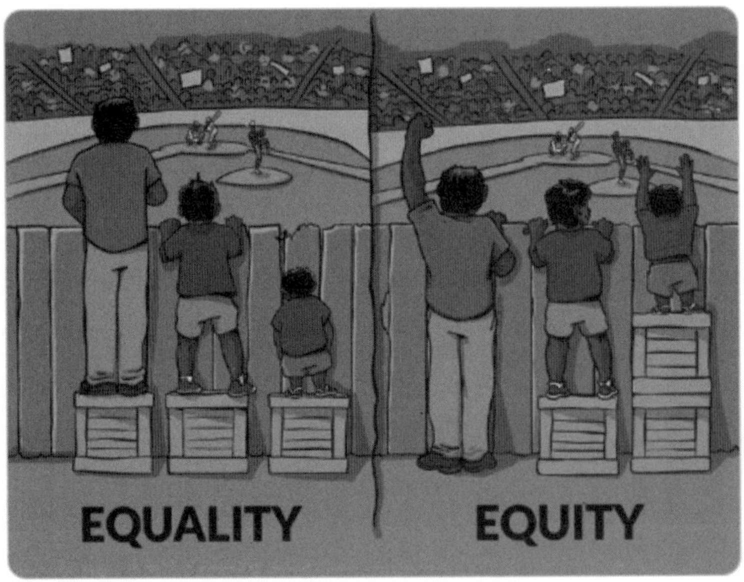

3:22 PM · Apr 26, 2019 · Twitter for iPhone

1.3K Retweets **7K** Likes

Tweet Link
https://twitter.com/anandmahindra/status/1121713774157688832

1. A Level Playing Field

From the Curator's Lens

The Mahindra Group has created a Diversity Council which aims to build an inclusive culture and diversity in their workforce. This enables a safe, positive, and nurturing environment to attract, engage and retain talent of all ages, genders, nationalities and abilities.

The choice of the image is most apt as it illustrates a clear understanding of the concepts of equality and equity. By posting this tweet, an empathetic leader conveys a strong belief in these ideals and the commitment towards propagating them.

#Women Empowerment
@Being Anand Mahindra

anand mahindra
@anandmahindra

As the movement to give equality & empowerment to women intensifes, we must not forget to acknowledge & empower women farmers who are the unrecognized backbone of our agriculture. At @MahindraRise we're committed to this cause...Please do watch the video when you have a moment..

Prerna for Women @PrernaforWomen · Oct 15, 2018
On the occasion of #MahilaKisanDiwas, we'd like to acknowledge the contributions made by women farmers in Indian agriculture & take them a step closer towards prosperity.#PrernaForWomen #MainHoonKisan #RiseForGood @MahindraRise @anandmahindra @goenkapk @rajesh664 @TractorMahindra

2:36 PM · Oct 16, 2018 · Twitter for iPad

116 Retweets **580** Likes

Tweet Link
https://twitter.com/anandmahindra/status/1052123698025492480

2. ***Prerna***

From the Curator's Lens

'Prerna' is a Mahindra Group program which is committed to women empowerment. It has empowered these women by giving them access to gender-neutral farming equipment. The program also disseminates knowledge about farming techniques which reduce drudgery and yet increase productivity.

On the occasion of *Mahila Kisan Diwas* (National Women Farmers Day), this tweet is an appeal to appreciate women farmers and their contribution to the agricultural sector in India. It prompts the reader to recognize the active presence of India's female workforce which is otherwise neglected.

#Women Empowerment
@Being Anand Mahindra

anand mahindra
@anandmahindra

She showed great wisdom in choosing this brand! My compliments to the team for supporting her & facilitating her training...

> **MANISH ARORA** @ManishAro · Oct 31, 2018
> And here is another example of women empowerment !! 1st lady to get offical licence to operate a backhoe loader !! @anandmahindra @RajanWadhera1 @GoenkaPk @vinod_sahay

Pulling more than her weight

How a woman who started working late in life now moves the earth

8:15 AM · Nov 1, 2018 · Twitter for iPad

44 Retweets **357** Likes

Tweet Link
https://twitter.com/anandmahindra/status/1057826033615937536

3. It Can Be Done

In the news article in this tweet, Lekha Rajendran is seen breaking the mould as she successfully operates a **Mahindra Backhoe Loader**. This particular area of work has traditionally been and continues to be male-dominated.

As the Mahindra Group stands against gender-bias, it is most natural that they would support Ms. Rajendran's efforts. That Ms. Rajendran was featured in the media, reiterates the importance of the company's attitude and approach towards women empowerment. Retweeting the post bears witness to the spirit of the Mahindra Group.

#Women Empowerment @Being Anand Mahindra

anand mahindra
@anandmahindra

Kudos to my wife&daughters & the Verve team for choosing to break boundaries with their 'cover girl.' It's not just models & conventional celebrities who belong on the cover; It's all those women who are breaking boundaries in their professions...

Verve Magazine @vervemagazine · May 10, 2018
Our May issue, dedicated to the Indian design aesthetic, features Mumbai-based artist @sabeenu who innovatively creates a variety of shapes through quilling.

5:24 PM · May 10, 2018 · Twitter for iPhone

59 Retweets **497** Likes

Tweet Link
https://twitter.com/anandmahindra/status/994546123258892289?lang=en

4. Paper Tigress

From the Curator's Lens

Verve was launched in 1995 by Mrs. Anuradha Anand Mahindra. It is the first domestically-owned luxury lifestyle magazine with over 60,000 regular subscribers. Keeping to the trend of unconventional choices of featured personalities, Sabeena Karnik graces the cover of the May 2018 issue. She is a celebrated quilling artist whose artwork is appreciated and used by many recognized brands as décor and in promotion.

This tweet carries the emotions of a proud husband and father who is encouraging, and supportive towards the professional journey of his wife and daughters Divya and Aalika.

#Women Empowerment
@Being Anand Mahindra

anand mahindra
@anandmahindra

I've been helping to baby-sit my year old grandson this past week & it's brought home to me the stark reality of this image. I salute every working woman & acknowledge that their successes have required a much greater amount of effort than their male counterparts

12:15 PM · Feb 5, 2019 · Twitter for iPhone

11.2K Retweets **34.7K** Likes

Tweet Link
https://twitter.com/anandmahindra/status/1092675489586569216?lang=en

5. Skewed Competition

While the fight for gender equality at the workplace is going on in many countries, the division of labour, in terms of household chores, is seldom discussed. However, when Mr. Mahindra spent time babysitting his one-year-old grandson, he was struck by the challenges faced by all working women.

This prompted him to take to Twitter to salute all such women and the struggles with balancing their household and professional duties. Furthermore, it initiated a conversation among *twitterati* on gender equality and work-life balance. This tweet generated a flood of responses on Twitter in the form of thousands of replies, retweets, and likes. Further, it was mentioned in The Economic Times and on indiatimes.com.

#Women Empowerment @Being Anand Mahindra

anand mahindra
@anandmahindra

And it's now a decade since my mother departed. But children remember even the earliest memories of their parents. That's my elder sister & I in the lap of a young woman we salute this Mother's Day...

Tweet Link

https://twitter.com/anandmahindra/status/1127451520457486336

6. Celebrating Mother's Day

From the Curator's Lens

A leader travels the distance from 'Good to Great' when imbibed with the qualities of gratitude and sensitivity. The image of a smiling mother holding her two happy children conjures up a sense of joy, security and family bonding.

Indira Mahindra, Mr. Mahindra's mother, was inclined towards the creative arts and literature. She has authored four books and taught history at the Allahabad University. Mr. Mahindra freely expresses an emotional connect with his mother in this tweet. He also tweeted the same warmth in an earlier tweet saying, "Mothers are forever, aren't they?"

#Women Empowerment
#Super Six Stats

	A Level Playing Field	*Prerna*	It Can Be Done
Retweets	1360	116	44
Likes	7029	584	359
Comments	217	35	14

	Paper Tigress	Skewed Competition	Celebrating Mother's Day
Retweets	59	11200	1279
Likes	500	34800	24217
Comments	19	1100	504

One's cultural identity is the holistic combination of
Family, Religious, Ethnic, Geographical, National, and Political factors.

@Being Anand Mahindra

#Culture & Heritage

संस्कृतिश्च परंपरा च।

स्थापत्यं रचना च शिल्पसुभगं संगीतनृत्यांगनम्
शांतिक्रांतिपरंपरागुणमयं स्थानं प्रभोः मंदिरम्
श्रद्धास्नेहगुणैर्निबद्धहृदयं दीपेन प्रज्वालितम्
मूल्यानामनुकर्षणेन सकलं विश्वं कुटुंबं भवेत्

"The temple of God and the heart are similar and so this description applies to both. Both show great architecture, structure, beautiful sculpture, music and dance. It is full of the light of a lamp. The heart has the qualities of peace, revolution and tradition and is bound with the strings or qualities of faith and love. If these values are cherished and followed, the whole world will be a family."

#Culture & Heritage

The Essence of **#Culture & Heritage - By the Curator**

Anand Mahindra epitomizes a global leader infused with an Indian ethos.

Heritage stands for something that is inherited and passed down from previous generations. Cultural heritage is thus the passing on of culture, values and traditions and implies a shared bond, a belonging to a community. It encompasses a wide range of tangibles such as paintings, drawings, prints, mosaics, sculptures, historical monuments and buildings; and abstract ideas such as traditions, oral narratives, performing arts, social practices, traditional craftsmanship, representations, rituals, knowledge and skills transmitted from generation to generation within a community.

Mr. Mahindra has a high regard for traditions and culture, be it Indian or global. He has been an influential advocate of the Indian value system.

In this era of globalization, cultural heritage helps us remember our cultural diversity, the understanding of which develops mutual respect and renewed dialogue amongst different cultures. Inter-personal connections are the foundations for social change. Building relationships with people from different cultures, is the key to creating diverse communities. The psyche of a person with high emotional quotient is able to be highly culturally sensitive. This is true of Mr. Mahindra as well.

Theory Model for Culture & Heritage

After studying 16,000 employees of IBM, Prof. Geert Hofstede concluded that there are two contrasting cultural orientations: one which values individualism, while the other which values collectivism. According to him, culture can be understood over four dimensions, namely Power Distance, Individualism, Uncertainty Avoidance and Masculinity.

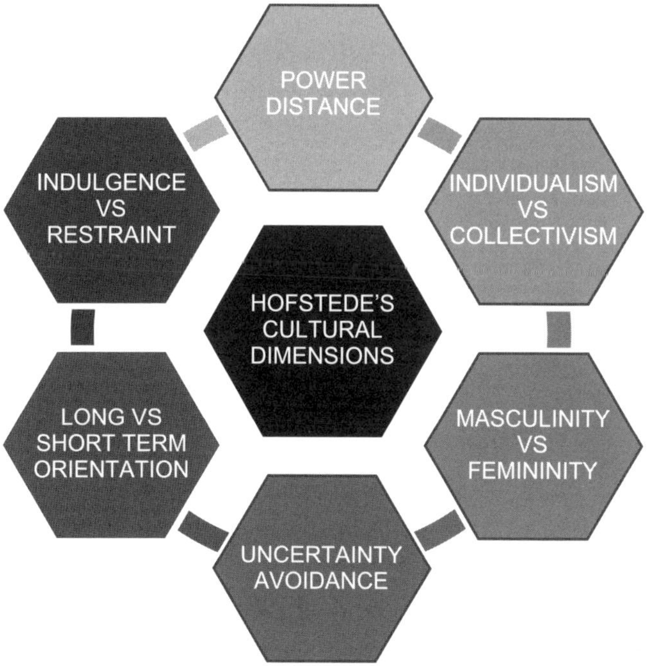

Geert Hofstede's Inter-cultural Dimensions Model [1]

Assessing Indian society as per the Hofstede model shows that its individualistic traits blended with a religious philosophy, reflect collectivism of its members. India is very 'masculine' in terms of visual display of success and power. It is an ancient and spiritual country with one of the longest surviving cultures. The culture has demonstrated the values of humility and abstinence; and displays a high tolerance for the unexpected.

With globalization , the workforce has become multi-cultural. Leaders need to be adaptive for managing such a culturally diverse workforce across international boundaries. In a career spanning 34 years with his business group, Mr. Mahindra has demonstrated a deep sense of purpose in building an enduring organization while staying true to his core values.

The Mahindra Group is a multi-national conglomerate having its operations in over 100 countries. The Group's culture is a perfect fit with the Hofstede model. Although the conglomerate has a global presence, it has been able to retain its Indian ethos. The culture of any organization depends on the attitude and philosophy of its leader.

The values and core belief of the Mahindra Group focus on the fairness and dignity of an individual. The company takes pride in saying that they provide equal opportunity and inclusion for all employees through their employment policies and practices.

Mr. Mahindra's tweets, speeches and writings are a true mirror reflecting his pride for Indian culture. He has expressed his thoughts about different facets of Indian culture like language, ethnicity, gender identity and nationalism. Remarkably, while he has maintained an Indian ethos, he has also been able to develop a global perspective.

An illustration of his pride in Indian culture is the tweet about regional food which is traditionally served in a platter or *thali*.[2] Similarly, a tweet, the one about being a 'renaissance' leader with cultural intelligence, he collaborates with foreign counterparts to leverage their synergies all the while keeping their cultural sensitivities intact. This has been evident from the different mergers & acquisitions, Joint Ventures the company has had with foreign companies like Pininfarina, JAWA, Holiday Club, Peugeot Motocycles, Mitsubishi Agri, Ford, CIE Automotive,[3] and Ssangyong.[4]

Mr. Mahindra takes pride in being an Indian and strives to enthuse a nationalistic spirit in all those around him. Remarkably, he does this in a non-aggressive way. He chooses to tweet about various festivals, iconic monuments and traditions by which he keeps cultural issues at the forefront. In his addresses, he often refers to ancient Hindu mythological figures as at the launch of the Xylo SUV at Nashik, when he mentioned the story Garuda and the foundation of the city.

A large number of Mr. Mahindra's tweets prove his inclusive approach which is derived from the concept of collectivism in Indian culture. It is a direct derivative of the concept of *Vasudhaiva Kutumbakam* (the whole world is one family) which is at the core of Indian culture.

Super Six Parameters

1. Institution	4. Values and Art
2. Principles	5. Tradition and Ritual
3. Beliefs and Practice	6. Heritage and Identity

1. https://www.hofstede-insights.com/country-comparison/india/
2. https://twitter.com/search?q=thali%20(from%3Aanandmahindra)&src=typed_query
3. https://twitter.com/anandmahindra/status/34652710033909760
4. https://twitter.com/anandmahindra/status/21132476228

#Culture & Heritage
@Being Anand Mahindra
Super Six

1

The Rise of a Geo-political Giant

2

Vasudhaiva Kutumbakam

3

The Ancient Indian Science of Well-being

4

A Solitary Wish

5

A Perfect Blend

6

Cultural Consciousness

#Culture & Heritage
@Being Anand Mahindra

anand mahindra
@anandmahindra

There'll be many opinions about the importance given to a Hindu temple.I look on the UAE's donation of land as a surprising&significant gesture which signals India's growing geopolitical importance in the region. Well done @navdeepsuri for enabling this during the PM's visit

Raveesh Kumar @MEAIndia · Feb 11, 2018
This is how the first Hindu temple in Abu Dhabi would look like once completed! PM @narendramodi unveiled the model of the temple and thanked the Crown Prince of Abu Dhabi for gracious donation of the land. The temple will be a testimony to UAE's commitment to tolerance & harmony

4:22 PM · Feb 11, 2018 · Twitter for iPad

400 Retweets **1.7K** Likes

Tweet Link
https://twitter.com/anandmahindra/status/962640366578098176?lang=en

1. The Rise of a Geo-political Giant From the Curator's Lens

The Prime Minister of India, Narendra Modi, came to power with a 'neighbourhood first' agenda. He strengthened both the socio-cultural as well as the political ties of India with these and many other nations. He connected with the Indian diaspora there in whom he invoked a sense of pride about Indian culture, religion and spiritualism.

The tweet turns the spotlight onto the transformational leadership qualities of Mr. Mahindra as it illuminates his passion for the Indian ethos. It also magnifies his strong nationalistic pride and desire to see India succeed at the global levels.

#Culture & Heritage
@Being Anand Mahindra

anand mahindra
@anandmahindra

Vasudhaiva Kutumbakam in action. I had no idea about it & I'm sure I'm not the only one. These seemingly small but important narratives of our collective culture should have been focused on in our history books much more than the great battles of our history

6:02 PM · May 21, 2018 · Twitter for Android

1.5K Retweets **3.1K** Likes

Tweet Link

https://twitter.com/anandmahindra/status/998542094460489728

2. *Vasudhaiva Kutumbakam*

Here unfolds a neglected page from history which every Indian should be aware of. The tweet proudly refers to a story underlining the concept of generosity, tolerance and compassion which is the essence of *Vasudhaiva Kutumbakam.*

Indians magnanimously welcomed and accepted the Polish Jews, just as they did Jewish refugees from other parts of Europe during and after World War II. Similarly, Parsi refugees from Western Asia were able to assimilate into Indian society. Noticing this forgotten piece of history and sharing it with pride, is the mark of a leader who proudly displays his Indian DNA.

#Culture & Heritage
@Being Anand Mahindra

anand mahindra
@anandmahindra

Proud that India's strongest 'weapon' of global influence isn't a nuclear arsenal but the Power of Pranayam & Yoga. Bravo @PMOIndia

Gaurav Sharma(高和) @tsgaurav · Jun 20, 2017
Around 150 Chinese and Indians perform #Yoga at the Great Wall of #China #InternationalYogaDay

10:12 AM · Jun 21, 2017 · Twitter for Android

776 Retweets **2.4K** Likes

Tweet Link
https://twitter.com/anandmahindra/status/877386168597336064

3. The Ancient Indian Science of Well-being

From the Curator's Lens

The age-old tradition of Yoga is the exercise of disciplining both the physical body and the meta-physical mind. India's unique gift to the world has been validated by the United Nations' declaration of June 21st as International Yoga Day. Yoga surpasses all divides such as race, religion, and region and is only concerned with humanity's endeavour towards leading a wholesome life.

Tweeting photographs of yoga enthusiasts in China on this day, Mr Mahindra exudes appreciation in seeing the Prime Minister's efforts towards popularizing yoga around the world. Furthermore, a sense of hope emanates from a global business leader that this exercise would prove to be more potent than any armed conflict to influence the world.

#Culture & Heritage
@Being Anand Mahindra

anand mahindra
@anandmahindra

As Diwali draws to a close, I have no colourful animated images to share. The image of a solitary Diya is my enduring memory of this great festival. One single lamp is all we need to maintain hope in our lives & bring hope to others. I wish you all good fortune & great happiness

8:51 PM · Nov 7, 2018 · Twitter for iPhone

436 Retweets **5.3K** Likes

Tweet Link
https://twitter.com/anandmahindra/status/994546123258892289?lang=en

4. A Solitary Wish

From the Curator's Lens

Diwali is the festival of lights which coincides with the Hindu New Year and celebrates new beginnings. The alternate word *deepavali* literally means a 'row of lights' which is traditionally meant to illuminate a path. Diwali is perhaps the most awaited of all festivals of India, as it brings delight and enjoyment to millions.

The conscious use of this particular image and caption reflects a deep understanding of the real significance of this festival. Only an inspirational leader would feel compelled to motivate others to the tune of 'less is more'. The image of a *diya* has a deeper meaning as it can only be lighted from one flame to another – *jyot se jyot jalate chalo.*

#Culture & Heritage
@Being Anand Mahindra

anand mahindra
@anandmahindra

The @MahindraRise Group just opened its first office in Washington D.C at the corner of New York Avenue and 14th Street. We're proud of our history and track record in the U.S and we will work to keep those flags flying...

7:03 PM · Jun 7, 2019 · Twitter for iPhone

1.4K Retweets **14.5K** Likes

Tweet Link
https://twitter.com/anandmahindra/status/1136989639305322498

5. A Perfect Blend

From the Curator's Lens

A moment of pride as a new Mahindra office is opened overseas. Here seen is the true ethos of the Mahindra Group as combining the traditions with the modern practices. They have set themselves up as a role model for other Indian companies to spread their wings globally, all the while keeping their traditions intact.

This echoes the fact that one can take Indians out of India, but not India out of Indians. In one of the accompanying photographs we see men in western suits sitting on the floor in traditional Indian style. This focuses attention on the seamless blending of corporate and traditional culture for which the company is well known.

#Culture & Heritage
@Being Anand Mahindra

anand mahindra
@anandmahindra

Interesting how a building/monument can become a symbol of national/personal as well as global identity for citizens of a country or city. I suppose for India the equivalent icon would be the Taj Mahal? What structure holds that status in your city?

BBC News (World) @BBCWorld · Apr 16

"If there's one site that symbolises the permanence of the French nation it is Notre-Dame"

BBC journalist @HenriAstier explains why Notre-Dame has remained a powerful symbol for Parisians

10:38 AM · Apr 17, 2019 · Twitter for iPhone

160 Retweets **1.5K** Likes

Tweet Link
https://twitter.com/anandmahindra/status/1118380643770093573

6. Cultural Consciousness

From the Curator's Lens

The tweet relates to a fire at the World Heritage site of the Notre Dame in France. Aware of the feeling of possessiveness one has about one's own culture, a question is posed for the *twitterati* to identify any famous monument close to their heart.

This highlights an awareness that such iconic imagery invokes a sense of regional and national pride. Only an individual with a high EQ and a strong cultural identity would be touched by this sort of incidence from another country. Additionally, the tone in this tweet urges readers to connect to their roots and motivates them to preserve their heritage.

#Culture & Heritage
#Super Six Stats

	The Rise of a Geo-political Giant	*Vasudhaiva Kutumbakam*	The Ancient Indian Science of Well-being
Retweets	400	1492	778
Likes	1662	3163	2419
Comments	38	143	74

	A Solitary Wish	A Perfect Blend	Cultural Consciousness
Retweets	440	1400	160
Likes	5330	14595	1495
Comments	225	846	711

"It is not the strongest of the species that survives, nor the most intelligent that survives. It is the one that is the most adaptable to change"

- Charles Darwin, English Evolutionist and Geologist

@Being Anand Mahindra

#Socio-Economic-Political Scenario

अर्थराजनयसमाजक्षेत्राणां परिदृश्यम् ।

अर्थस्याकलनं समाजपरता उद्योगतंत्रज्ञता
व्यापारे शुचिता विधायकदले संघातविज्ञानता
आचारे चलता नियोजनमये कार्ये स्वयंप्रज्ञता
एतासां सकलात् सुखानुसरणात् वैश्वं समाराधनम्

“If the qualities of comprehension of the economy, connecting with society, knowledge of business and technology, honesty in trade, skill of organizing of different communities, consistency in conduct, innovation and planning in the work culture are easily and totally adopted, it works successfully for the satisfaction of the whole world.”

#Socio-Economic-Political Scenario

The Essence of

#Socio-Economic-Political Scenario - By the Curator

Anand Mahindra has been a keen observer of all the factors that influence business strategies and have a long-term impact on society. His tweets connote a thorough understanding of PEST factors and their implications.

There are various Political, Economic, Social and Technological (PEST) factors that are constantly influencing an organization's performance. PEST analysis is one of the most widely used tools for analyzing a business environment. It is a way of understanding how external forces impact one's business.

While commenting on any political phenomena, Mr. Mahindra refers to political and policy factors that influence business decisions. In his comments on the economy & economics, he examines the issues that can play a role in the success of both his company as well as the country. He looks at social events to understand how consumer needs are shaped. He understands that technology plays a huge part in business as it can have either a negative or positive impact.

A visionary business leader identifies present and future threats and opportunities associated with the plethora of factors. Mr. Mahindra is such a leader.

Theory Model for Socio-Economic-Political Scenario

Analysis of the business environment is the systematic process by which factors that affect any business are identified, their impact is assessed and a strategy is developed to mitigate and/or take advantage of them. One of the most widely used tools for conducting such an analysis is PEST.[1] This tool is also relevant here as it helps relate to Mr. Mahindra's business and marketing strategies.

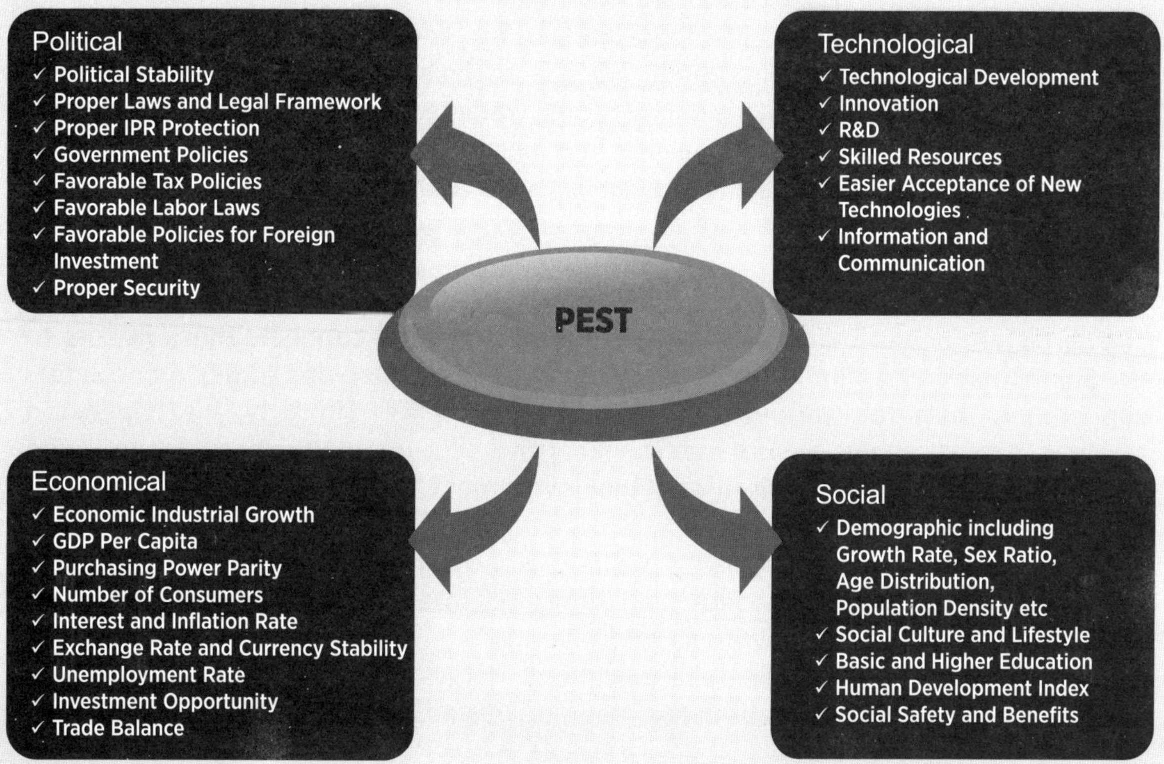

PEST
Business Analysis

Many of the tweets by Mr. Mahindra reflect his keen understanding of factors influencing the business environment.[2] He has analysed and commented on a wide range of issues such as international relations, trade, economic policies and political incidences. Each of these factors has implications for individuals, businesses and the society at large. They define the present as well as future course of action of an organization and therefore, need to be analysed in a larger context.

While commenting on economic factors, Mr. Mahindra uses fundamental economic concepts, jargon and theoretical models. Furthermore, he connects these factors to various national and international scenarios reflecting inter-dependence and inter-connectivity between various economic developments. Similarly, his tweets on political events show an optimistic, receptive, futuristic business leader who understands the vision and priorities of the government.

As an advocate of gender equality, women empowerment and social institutions, Mr. Mahindra assertively tweets on such issues. His interest in technology is mirrored via the tweets on local innovations and inventions and international technological advancements and developmental implications. His tweets reflect a deep theoretical and applied knowledge base about the changing business environment.

Mr. Mahindra's undisputable ability to evaluate and scrutinise the changing business environment and his passion to share the same with all, is an important attribute that makes him a transformational leader.

1.https://www.mindtools.com/pages/article/newTMC_09.htm
2.https://accountlearning.com/6-important-factors-business-environment-influence-business/

PEST Model image source:
https://carpenterstrategytoolbox.com/2016/01/20/add-a-pest-to-your-analysis/

#Socio-Economic-Political Scenario
@Being Anand Mahindra

Super Six

1 Strategic Economic Partnership

2 Cryptic Currency

3 Envisioning National Growth

4

Foreseeing Economic Development

5

Enhancing Connectivity

6

Transforming the Skies

#Socio-Economic-Political Scenario @Being Anand Mahindra

anand mahindra
@anandmahindra

A sign of things to come. For developing economies, the combination of India & Japan provides a powerful alternative to China's embrace. Expect to see an acceleration of this alliance in @narendramodi 's new regime.

> **BloombergQuint** @BloombergQuint · May 21
> India, Japan and Sri Lanka are set to sign a MoU to build out the east container terminal at the Port of Colombo.
> bit.ly/2WV88UX

4:38 PM · May 30, 2019 · Twitter for iPhone

429 Retweets **2.7K** Likes

Tweet Link

https://twitter.com/i/web/status/1134053982136086528

1. Strategic Economic Partnership

From the Curator's Lens

The tweet highlights India's changing foreign policy outlook in Indian Prime Minister Narendra Modi's new regime. It clearly indicates how PM Modi has redefined strategic autonomy as an objective that is attainable through strengthened economic partnerships.

Mr. Mahindra's pan-sectoral understanding is revealed via this comment which relates to a multi-national port sector undertaking involving India with Sri Lanka and China. It emphasises the broad outlook and understanding of this inspirational leader which warranted his attention and prompted a tweet.

#Socio-Economic-Political Scenario @Being Anand Mahindra

anand mahindra
@anandmahindra

Good move. My friend Arjun Divecha,a renowned fund manager, had the best comment on cryptos: They're like feathers in a storm; the gusts of wind blow them higher,but at some point they hit the ground.

Govt committee recommends ban on cryptocurrency in India
In a setback for cryptocurrency enthusiasts and start ups, India seems all set to ban private cryptocurrencies after an inter-ministerial committee (IMC) suggeste...
indiatoday.in

2:56 PM · Jul 23, 2019 · Twitter for iPad

199 Retweets **1.4K** Likes

Tweet Link

https://twitter.com/anandmahindra/status/1153597233306058753

2. Cryptic Currency

From the Curator's Lens

The Indian government appointed an inter-ministerial committee for reviewing and scrutinizing the pros and cons of cryptocurrency in July 2019. It recommended a ban on all cryptos except the ones issued by the state. This move has been appreciated by many on account of the risks involved in cryptocurrencies.

Through this tweet Mr. Mahindra supports the committee's recommendations and warns of the drawbacks involved in the use of cryptocurrency. He paraphrases his friend when stating that the weightlessness of a feather illustrates the extreme volatility of the value of cryptocurrency.

#Socio-Economic-Political Scenario @Being Anand Mahindra

anand mahindra
@anandmahindra

Excellent. I had just tweeted about the need to roll up our sleeves & get the GDP growth engine moving faster. Clearly, the RBI is leading from the front.. 👏👏👏 @DasShaktikanta

Economic Times @EconomicTimes · Aug 7
#BREAKING | @RBI cuts repo rate by 35 bps to 5.40%, maintains accommodative stance
#RBIMonetaryPolicy #RBIPolicy bit.ly/2YvWPHu
Show this thread

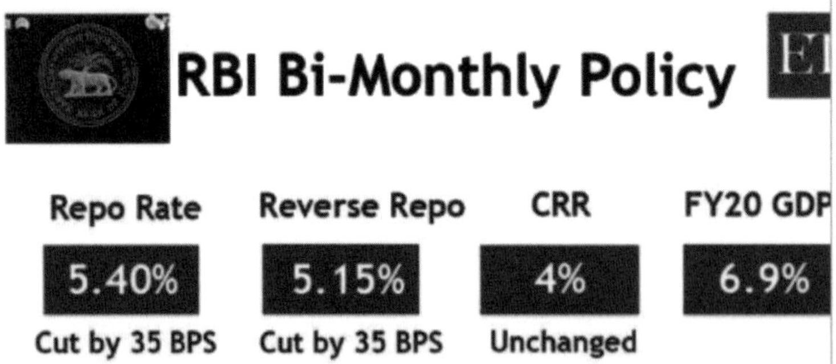

12:30 PM · Aug 7, 2019 · Twitter for iPhone

395 Retweets **3.5K** Likes

Tweet Link
https://twitter.com/anandmahindra/status/1158996330217234432

3. Envisioning National Growth

From the Curator's Lens

The tweet throws light on the relationship between monetary policy and the growth of GDP. During a global turndown, a lower Repo rate implies cheaper credit availability for the banking sector which, in turn, is passed on to the retail credit customers of commercial banks. It is further expected that this serves to boost private consumption and improve investment activity.

It is no wonder that Mr. Mahindra feels excited by this news in The Economic Times. Any industrialist who has invested in the growth of the nation would think on the same lines. This is why the policy rate cut under the leadership of the Reserve Bank of India Governor, Shaktikanta Das, has won praise here.

#Socio-Economic-Political Scenario @Being Anand Mahindra

anand mahindra
@anandmahindra

Our Group Economics Division produces this compact bulletin at the start of every week. Instead of a post mortem of major economics events we wanted advance notice of key developments to look out for. Will share it on twitter if people are interested...

Mahindra *Rise.*

#Things this week

Brexit: One more vote? (Tue – Mar 19th?): Theresa May warned that unless they approved her Brexit divorce deal, Britain's exit from the EU could face a long delay. Her deal, an attempt to keep close relations with the EU while leaving the bloc's formal structures, was defeated by 230 votes in parliament on Jan. 15 and by 149 votes on March 12.

But May continues to fight to build support for her plan, which is expected to be put before lawmakers for a 3rd time, possibly on Tuesday.

To get it through parliament, she must win over dozens of Brexit-supporting rebels in her own Conservative Party and the Northern Irish Democratic Unionist Party which supports her minority government.

Chancellor of the exchequer Philip Hammond also increased pressure on Brexiteer Tory MPs, warning they will trigger a long delay to leaving the EU unless they support May's proposed deal when it returns to Parliament.

What to watch out for:

A growing number of Tories are now backing her agreement, Hammond said, and **the deal will only be put to another vote this week if the government is confident of victory**. The British Pound has largely remained unchanged and will continue to be a lead indicator.

US Fed (Wed, Mar 20th): **In a policy-heavy week, 7 central banks will announce their monetary policies** including that of the US, UK, Taiwan, Thailand, Malaysia, Philippines and Indonesia. However, all eyes will be on the US Federal Reserve, which is expected to announce its policy on Wednesday at the end of its two-day policy meeting.

What to watch out for:

The Fed Funds rate, which is in a range between 2.25%-2.5% is likely to remain unchanged.

While the Fed policy is broadly expected to be on hold, with the 'patient' stance, the markets would have their eyes glued on the Fed's language. Fed Chair **Jerome Powell's post policy interaction will be a closely-watched thing**.

The **Fed will also release new forecasts for US economic growth and interest rates, known as the "dot-plot"**, which will probably point to either no more rate hikes this year or to one more at most.

The **Fed will also likely announce further details on when it will end its operation to unwind its balance sheet**.

Tweet Link
https://twitter.com/anandmahindra/status/1107690320106385408

4. Foreseeing Economic Development

From the Curator's Lens

A visionary leader understands the importance of analysing and sharing the information about important economic developments. As such, sharing the unique bulletin published by Mahindra Group's Chief Economist is the natural result of this way of thinking.

The contemporary global economic scenario is in the volatile, transformation phase. At the same time, international economic power centres have begun to shift from the West to the East. These factors will have long-lasting repercussions for businessmen, policy makers and the society at large. Consequently, understanding these changes and challenges is necessary for making rational decisions. The Mahindra Group Economic Division Bulletin provides advanced analyses of anticipatory changes and will certainly prove timely regarding any strategic move in business and industry.

#Socio-Economic-Political Scenario @Being Anand Mahindra

anand mahindra
@anandmahindra

Bhavnagar to Bharuch 350 kms by road 32 kms by sea route. Transport using waterways is becoming a reality. Step by step, the logistics of transport, travel & trade in India is becoming more efficient.

1:27 PM · Nov 29, 2018 · Twitter for iPhone

7.5K Retweets **24.6K** Likes

Tweet Link

https://twitter.com/anandmahindra/status/1068051249206099968?lang=en

5. Enhancing Connectivity

From the Curator's Lens

South Asia's first world-class ferry, Phase 1 of the 'Roll-on, Roll-off (Ro-Ro)' service between Ghogha and Dahej in Gujarat was inaugurated by Indian Prime Minister Narendra Modi in October 2017. A Ro-Ro ferry is a vessel which carries wheeled cargo (cars, trucks, semi-trailer trucks, trailers, and railroad cars) that are rolled on and off the ferry on their own wheels or using a platform vehicle.

While commenting on the video of this event, Mr. Mahindra reiterates the importance of developing waterways as an efficient logistics service in India. Water transportation is deemed as the cheapest and fastest of all modes of transport. Further, this service will also save fuel and help decongest roadways across the country. The tweet appreciates this important infrastructural initiative by the Indian government.

#Socio-Economic-Political Scenario @Being Anand Mahindra

anand mahindra
@anandmahindra

All the focus is on Electric vehicles at the 'ground' level. Time to turn one's attention to the skies because electric aviation may well steal a lead on cars. And I'm not surprised that the tiny, innovative nation of Israel is in the forefront.

6:34 PM · Aug 2, 2019 · Twitter for iPhone

2.5K Retweets **10.8K** Likes

Tweet Link
https://twitter.com/anandmahindra/status/1157276087291801600

6. Transforming the Skies

From the Curator's Lens

The Israeli start-up, Eviation, showcased an aircraft prototype at the Paris Air Show in June 2019. This was their electric commuter plane called Alice. Developers hope to eventually increase its size and range. Mr. Mahindra appreciates Israeli innovation and enterprise overall. A leader with an eye for Innovation, leads the Mahindra Group towards becoming a major player in the electric automobile market in India.

The electric automobile market in India is expected to reach nearly $2 billion by FY2023 due to growing concerns about air pollution for the consumers, automobile companies and the government. The tweet with the video of the Eviation e-plane encourages Indian minds to focus on the potential market in electric aviation.

#Socio-Economic-Political Scenario #Super Six Stats

	Strategic Economic Partnership	Cryptic Currency	Envisioning National Growth
Retweets	436	202	407
Likes	2696	1383	3543
Comments	26	301	84

	Foreseeing Economic Development	Enhancing Connectivity	Transforming the Skies
Retweets	116	7549	2539
Likes	1225	24767	10835
Comments	175	492	305

"Behavioral Intelligence describes someone who not only possesses Emotional Intelligence,

but knows when, how, with whom and to what degree to apply it"

Sarah Canaday
Leadership Development Expert

@Being Anand Mahindra

#Current Happenings

वर्तमानकक्षे संवृत्तयः।

राष्ट्रेऽस्मिन्महिमा समाजसमता ध्येयं प्रजाशासनम्
व्यापारे पटुता कलानिपुणता सृष्टेश्च संपन्नता
तत्त्वज्ञानयुता शिवे भजनता प्राचीनपुण्यात्मता
एतत्सर्वशुभंकरं गुरुपदं त्वय्यस्ति भो भारत

"O Nation of India, you possess greatness, equality in the society, the ideal of democracy, expertise in business, proficiency in arts, glory of nature, devotion for the god of holiness with philosophy and the great merit of sacredness of ancient heritage. You have the place of the guru of the world with all these auspicious qualities."

#Current Happenings

The Essence of
#Current Happenings - By the Curator

One of the vital traits of an inspirational leader is alertness. This alertness is dependent on his personal attitude, perception and world view. An impactful leader possesses not only a high IQ (Intelligence Quotient), but a high EQ (Emotional Quotient) as well. This leader invariably responds to myriad happenings around him with great competence. His high speed of thought in recognizing changing circumstances and acting upon them makes him extraordinary.

The tweets selected reflect Mr. Mahindra's understanding in areas across the spectrum including business, economy, politics, nationalism, sports and cinema. Further more, his tweets have a value-addition of videos, photos, and hashtags. The content of these posts is multi-layered, rich and insightful. They motivate the readers to look beyond themselves and broaden their perspective towards life.

Mr. Mahindra is successful in relaying new tenets which stimulate people to grow intellectually and in compassion.

Theory Model for Current Happenings

The Entrepreneurial Intention Model illustrates the important aspects of an influential leader. An entrepreneur's knowledge is a product of his personality and experience which influences his entrepreneurial intention. Also, the perceptive ability of a leader is related to his level of alertness.

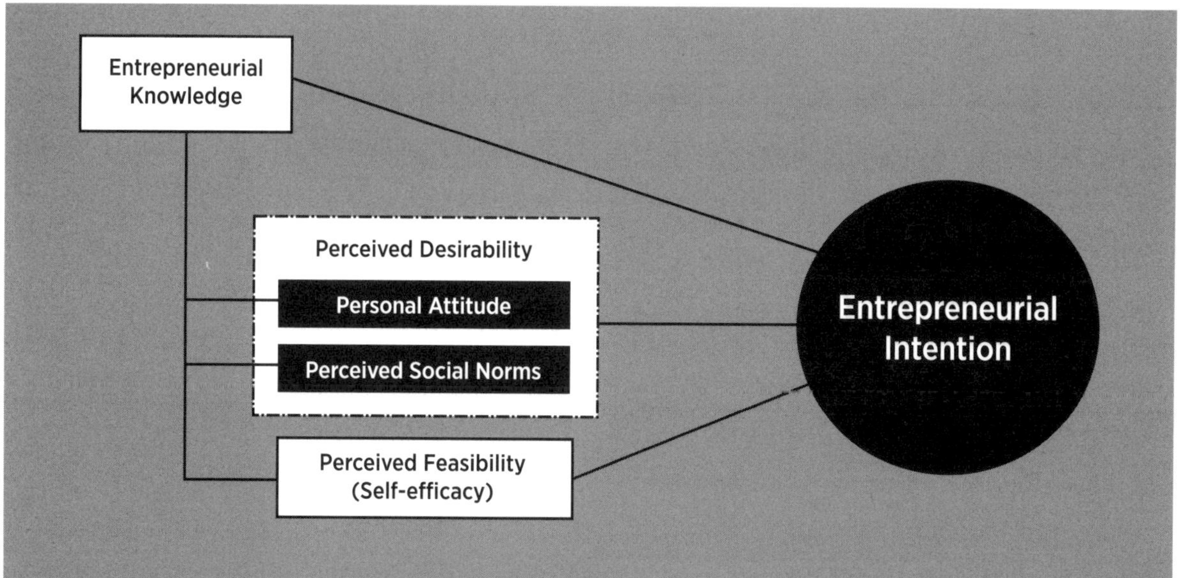

Entrepreneurial Intention Model[1]

A good leader with self-determination acquires the above quality. Leaders with great alertness have the inherent capacity to respond to the events happening around them. Such individuals tend to have a shorter reaction time, in which they not only think, but process the situation and act accordingly. A visionary leader can easily mentally shift from the past to the present, and on to the future and so recognizes any potential challenges before they become problems.

Mr. Mahindra, being an empathetic and attentive leader, is able to connect with various strata of society. He regularly tweets about a variety of incidences which he experiences and also retweets significant incidences that he comes across.

He was quick to connect to Dakshinmurthy Krishna Kumar's story (reminiscent of *Shravanbaal*, an Indian mythological figure famous for his devotion to his parents) and gift him an SUV.[2] By such actions, Mr. Mahindra brings to light a distinct perspective of looking at the world.

Mr. Mahindra's philosophy drives the Mahindra Group which has a strong community service portfolio under the umbrella of "Rise For Good" under the key focus areas - Girl Child Education, Youth Skilling and Farmer Prosperity. When the landmark verdict of the Ayodhya Ram Mandir[3] case was delivered by the Supreme Court of India, he showed respectful restraint in appreciating the judgement. Such tweets are a testimony to Mr. Mahindra's proactive interest in the well-being of his country and his fellow Indians. He continually strives to pique the interest and raise the awareness of the readers towards an even better India.

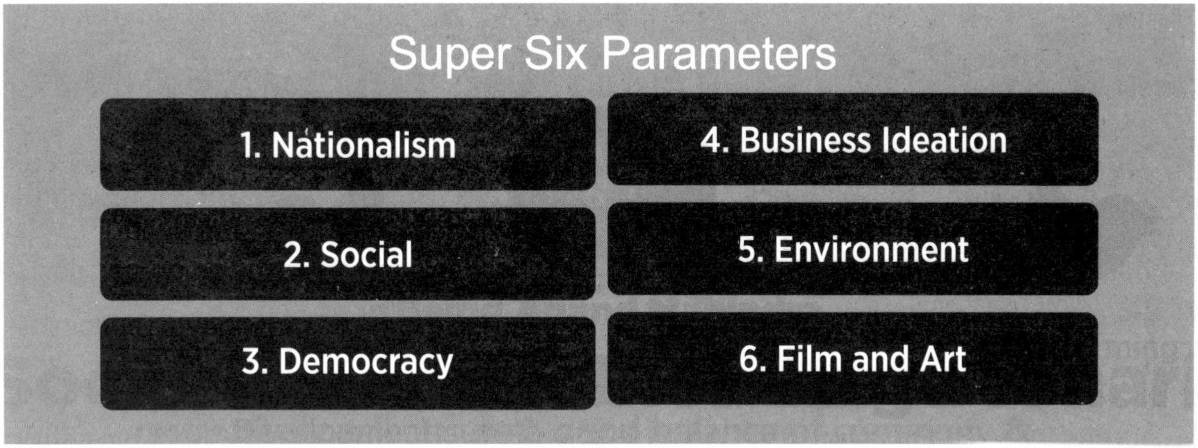

1. https://www.researchgate.net/figure/Entrepreneurial-intention-model_fig1_23731650
2. https://twitter.com/anandmahindra/status/1186870506529542144?lang=en
3. https://twitter.com/anandmahindra/status/1193058875370426368

#Current Happenings
@Being Anand Mahindra

Super Six

1. Scientists- The New Celebrities

2. Compassion for Community

3. Good Governance

4 The Indore Model

5 The Sound of the Monsoon Rains

6 A Moment to Celebrate

#Current Happenings @Being Anand Mahindra

anand mahindra
@anandmahindra

Was waiting to see the @isro folks hug each other before breathing normally again! Validation of our cryogenic tech is a technological breakthrough. It will enable us to make many more moon missions. I stand & salute our scientists. They are deservedly our new celebrities 🙏

Times of India @timesofindia · Jul 22
#ISRO #ISROMissions

Indian Space Research Organisation Chief K Sivan and other scientists celebrate after #GSLVMkIII-M1 successfully injects #Chandrayaan2 spacecraft into earth orbit.

For live updates: timesofindia.indiatimes.com/india/isro-cha...
Show this thread

3:20 PM · Jul 22, 2019 · Twitter for iPad

1.9K Retweets **17.7K** Likes

Tweet Link

https://twitter.com/anandmahindra/status/1153240881144557574

1. Scientists - The New Celebrities

From the Curator's Lens

India's greatly anticipated lunar mission, Chandrayaan-2, was launched and equipped with the country's most powerful space vehicle fully designed and fabricated in India. This achievement has enhanced national pride and confidence in the Indian scientific community.

Recognizing the contribution of the ISRO scientists is the mark of an alert and attentive leader. Mr. Mahindra insists on calling them celebrities whereby transforming the definition of the word to become more inclusive. He has pioneered a change in the hitherto societal mindset of glorifying only entertainment or sports personalities.

national pride

#Current Happenings @Being Anand Mahindra

anand mahindra
@anandmahindra

They may not figure in the Forbes Rich list but in my view, they are amongst the richest people in our country.Their wealth is their attitude to life. The next time I'm in their town I am definitely dropping by for tea & a tour of their exhibits..

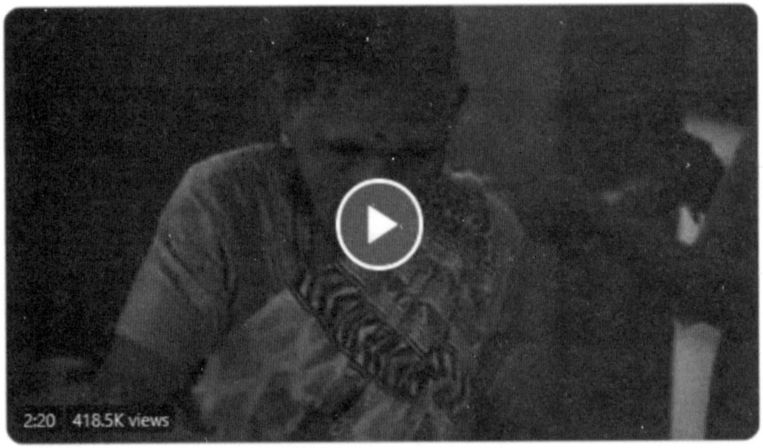

11:55 AM · Jan 9, 2019 · Twitter for iPhone

11.9K Retweets **33.2K** Likes

Tweet Link
https://twitter.com/anandmahindra/status/1082886138170228736

2. Compassion for Community

From the Curator's Lens

Wealth should not be counted just in terms of money. In the eyes of Mr. Mahindra, the couple referred to in this tweet is the perfect example to support this thought. The tweet further inspires the community by focusing attention towards those who are actually the cornerstones of society.

The tweet and accompanying video, revealing what one can accomplish by sheer will power, is no less than what can be achieved by wealth or power. Here is an instance of a humane leader's compassion for the dreams and aspirations of the common man. His appreciation for their accomplishments, in the form of their photographs and currency collection, is noteworthy.

#Current Happenings
@Being Anand Mahindra

anand mahindra
@anandmahindra

The ceremony started & finished right on time without any unnecessary additions to the program. Speed, focus & efficiency. If that's symbolic of how the new Govt will work, then it's time to be optimistic about the future.Congratulations to PM @narendramodi & his new team.

PMO India @PMOIndia · May 30
A team that is committed to fulfilling aspirations of 130 crore Indians.

Pictures of the Council of Ministers who took oath earlier today.

12:23 AM · May 31, 2019 · Twitter for iPhone

2.9K Retweets **20.9K** Likes

Tweet Link
https://twitter.com/anandmahindra/status/1134170871893913600

3. Good Governance

From the Curator's Lens

After the Prime Minister's Office of India shared pictures of the swearing-in ceremony in 2019, Mr. Mahindra tweeted this congratulatory message. He also took this opportunity to highlight the "speed, focus & efficiency" of the grand event. Furthermore, he express the hope that this would also be the style of working of the newly installed Modi cabinet.

The connect shown is very meaningful as it underlines the expectations of the government and the faith that they will be fulfilled. As an impact-maker, this method of backing the new government is very unique. When a leader, who is a steadfast nationalist, tweets in this manner, he also invokes in the reader a sense of pride in one's country.

#Current Happenings @Being Anand Mahindra

anand mahindra
@anandmahindra

One of our fastest-growing 'Rise' businesses—one that does well but also creates value for the community. We're proud to have created a template that will be emulated world-wide. This 'energises' us and makes us determined to grow this business even faster than originally planned

Som Narayan @SomNarayan12 · Aug 30
One more puri duniya se kaho copy that moment:) Congratulations Team Mahindra. Proud to be working with you all in Indore. UN passes resolution to adopt Indore model by 72 nations!! @anandmahindra @plpalani @subash_gp @KevinHouston2 @hsikka1 @MahindraLtd @carbonlites

72 nations to adopt Indore's waste-to-energy model

Resolution Passed By The United Nations

Times News Network

Indore: In Swachhta, Indore leads rest of the world follows. Following the footsteps of Swachh city, 72 Asian and African countries are now replicating the bio-methenation model for treatment of wet waste into bio-CNG.

A resolution in this regard was passed in a conference of

THE INDORE PROTOTYPE

ADVANTAGES

- Most sustainable zero waste model for wet waste management
- It fulfils the need of cost effective technologies of partner countries

CITY SET-UP

Two Units Already Operational

Choithram Mandi 8 tonne/day

Kabitkhedi 15 tonne/day

One Proposed At Trenching Ground 50 tonne/day

PLAN OF ACTION

- UN, with help of Japan govt, will install one plant of 50 tonne capacity in each country
- After that, it will be up to respective nations to carry on with initiative or develop it further

7:41 PM · Sep 3, 2019 · Twitter for iPhone

718 Retweets **3.2K** Likes

Tweet Link

https://twitter.com/anandmahindra/status/1168889291910647819

4. The Indore Model

From the Curator's Lens

The tweet refers to the Indore bio-methane plant of **Mahindra Rise** where approximately 35 tonnes of wet waste is treated and more than 1300kg bio-CNG is produced daily. This is a prime example of the implementation of the Mahindra Group's "Waste to Energy" model. The successful implementation of this innovative technology in India, prompted the United Nations to pass a resolution promoting this model worldwide and it has been used by 72 countries so far.

It is remarkable that a global corporate leader, along with profits, is concerned with environmental sustainability. Retweeting the news article appreciates the Rise team for achieving this milestone and motivates it towards speedier growth.

#Current Happenings @Being Anand Mahindra

anand mahindra
@anandmahindra

Finally. A steady patter of rain in Mumbai. Hard to explain to non-Indians what emotions the monsoon provokes in us. The poetry of anticipation & romance; the answer to our prayers; the promise of prosperity. Don't need anything else for getting my #MondayMojo going.

11:39 AM · Jun 24, 2019 · Twitter for iPhone

423 Retweets **5.7K** Likes

Tweet Link

https://twitter.com/anandmahindra/status/1143038300103565312

5. The Sound of the Monsoon Rains

From the Curator's Lens

Here, Mr. Mahindra has touched upon millions of Indian's deep connection with the advent of the monsoons. Ancient Indian scriptures, literature, art, sculpture, and music over the ages are replete with narratives and motifs of the monsoon rains. The Indian monsoon, while a basic necessity, has an overwhelming emotional connect. Over-riding all disparities and divides, Indians are united in praying for these yearly rains as they bring with them a promise of prosperity and well-being.

The tweet brings to the fore a person who crunches numbers and appreciates beauty equally ably. It echoes his empathy towards his fellow citizens whose very lives are dependent on this rain. He has a sensitivity which turns Monday Morning Blues into a Monday Mojo.

#Current Happenings
@Being Anand Mahindra

anand mahindra
@anandmahindra

Lookin' good, Team @BombayRoseFilm ! I can hear the applause all the way from Venice.... Thank you #GitanjaliRao, for making India proud with your exquisitely crafted film. And I'm proud that this is the 1st film for which I've taken the title of Co-producer. @cinestaanfilmco

CinestaanFilmCompany @cinestaanfilmco · Aug 29
Proud and delighted as our production, #BombayRose opens Venice International Critics Week to a tremendous response! #BiennaleCinema2019 #GitanjaliRao @anandmahindra #RohitKhattar @cinestaanfilmco @lesfilmsdici

11:13 AM · Aug 30, 2019 · Twitter for iPad

28 Retweets **429** Likes

Tweet Link
https://twitter.com/anandmahindra/status/1167311786724347905

6. A Moment to Celebrate

This tweet refers to **Bombay Rose** (2019), an internationally acclaimed animated film. It was nominated for Best Director at both the Chicago International and Venice Film Festivals in 2019. Mr. Mahindra acknowledges the excellence of the film by appreciating its animator and director, Gitanjali Rao and feels pride in having co-produced it.

That Mr. Mahindra can authoritatively tweet on this subject is testimony to his studies in the field of Film and Photography. A film enthusiast, he has taken to Twitter to express his educated views on other films as well. As a producer with a stake in an Indian entertainment company, it is a moment of great pride when his product headlines an international film festival.

#Current Happenings #Super Six Stats

	Scientists -The New Celebrities	Compassion for Community	Good Governance
Retweets	1939	12083	2957
Likes	17850	33614	21072
Comments	201	846	176

	The Indore Model	The Sound of the Monsoon Rains	A Moment to Celebrate
Retweets	723	430	29
Likes	3233	5794	428
Comments	92	170	17

Blue Tick

Pragraha, One who leads; inspires. When a person exudes leadership qualities which evolve to the highest level with experience and insight, he becomes a leader for all times.

Communication patterns have changed since the advent of social media. Today, micro-blogging and using images and/or video has proven an extremely effective means of reaching out to a large audience at once.

When the said leader excels at using the most modern methods of communicating, and it comes to the attention of an equally evolved academician, a Tweebook is born. However, the person who makes it an absolutely indispensable academic accessory, is the reader who studies it and grasps the knowledge to eventually spread it in the best possible manner.

This Tweebook is an unconventional medium of education for the futuristic generation who is attracted to the social media platform than conventional reading tools. As such, it is the hope that it will be valued as an interactive, learning tool.

Whether the printing medium be paper or digital, the words of the wise will remain eternal.

PS

Post-Script

In December 2019, headlines announced Mr. Anand Mahindra's decision to appoint new leadership roles at the end of FY2019.

Mr. Mahindra's conscious decision to hand the baton of running his conglomerate was well-received in print, digital and on social media. It is observed that he used democratic principles, rather than dynastic deliberations in the year-long process of choosing these leaders. It is notable that he valued professional capabilities above all other considerations while naming his successors.

I, along with my team, congratulate Mr. Mahindra on being conferred the ***PADMA BHUSHAN***, the prestigious Indian civilian award - 2020. At the conclusion of this epilogue, I feel elated that these recent decisions go a long way in reiterating Mr. Mahindra's Level 5 leadership skills as portrayed in the Tweebook. Let me wish him and his team sustained success in their professional endeavours which will continue to contribute to national growth.

This effort of presenting a *nishkaam karmayogi* (one who relentlessly toils without expectation of personal gain) is a salute to the incomparable and inimitable Mr. Anand Mahindra.

@twitterstatus

anand mahindra
@anandmahindra

Delighted to announce our leadership transition plan which reflects the Group's commitment to good governance. I am grateful to the board & nominations committee for conducting a diligent & rigorous year-long process that will ensure a seamless transition bit.ly/35BjotL

2:20 PM · Dec 20, 2019 · Twitter for iPad

https://twitter.com/anandmahindra/status/1207946423574024192

#twitterreactions

Pawan K Goenka
@GoenkaPk

Reflects M&M's depth of management talent. Having worked with all the new leaders, I am confident that the future of Mahindra is in good hands. @anishshah21 & @rajesh664 at the helm of the group & AFS respectively will ensure a good balance of continuity & change.Congratulations

https://twitter.com/GoenkaPk/status/1207948194719526919

Anish Shah
@anishshah21

Replying to @anandmahindra

Thank you @anandmahindra, I'm humbled by the trust reposed by you and the Board and honored to be a custodian of our values and culture. I'm excited about shaping our future, to drive growth and returns for our shareholders while enabling our communities to rise.

3:39 PM · Dec 20, 2019 · Twitter Web App

https://twitter.com/anishshah21/status/1207966223046078466

#twitterreactions

Rajesh Jejurikar
@rajesh664

My gratitude to the Board,@anandmahindra, @GoenkaPk for the opportunity to steer the Auto and Farm Sectors.I am humbled and excited,as we aspire to reach new heights by enabling our customers and communities to Rise

https://twitter.com/rajesh664/status/1207951160793018370

Parthasarathy V S
@parthavs

Getting back to leading businesses! Looking forward to the new role, heading the newly-created Mobility Services Sector at Mahindra. Proud of the structured and rigorous process followed by the Board. Thank you @anandmahindra. Congratulations to Leaders with new & expanded roles!

https://twitter.com/parthavs/status/1207982335318056960

#twitterreactions

Ruzbeh Irani
@RuzbehIrani1

People are the bedrock of any business. I feel excited to be given this opportunity to work closely with all the Mahindra Group businesses to lead and shape the future of HR in the Group.

https://twitter.com/RuzbehIrani1/status/1207948938398945280

CP Gurnani
@C_P_Gurnani

Thank you @anandmahindra & the board. Am honoured & humbled to be a part of this prestigious board.

I look forward to contributing to the growth of our company and continue to spread the mantra of Rise.. My heartiest wishes also to my colleagues on their well deserved new roles

https://twitter.com/C_P_Gurnani/status/1207995928067506176

Acknowledgements

Innovator & Curator	Dr. Prachee Javadekar
Design & Tech Credit	Dr. Aditee N. Vaidya
Research Team	Dr. Sonal Bhat Gandhali Divekar Dr. Sachin Bhide Dr. Girija Lagad Dr. Harshada Vaidya-Kannur Varsha Gajendragadkar Anuradha Deodhar Rupali Vakil
Sanskrit Shlokas Composed by	Dr. Supriya Sahasrabuddhe
Design & Development Team	Neha Agarwal Haria Ramola Dhande Payal Shah Karwa Kaushik Singh Surabhi Ghodgekar Sunil Suryawanshi